D0077047

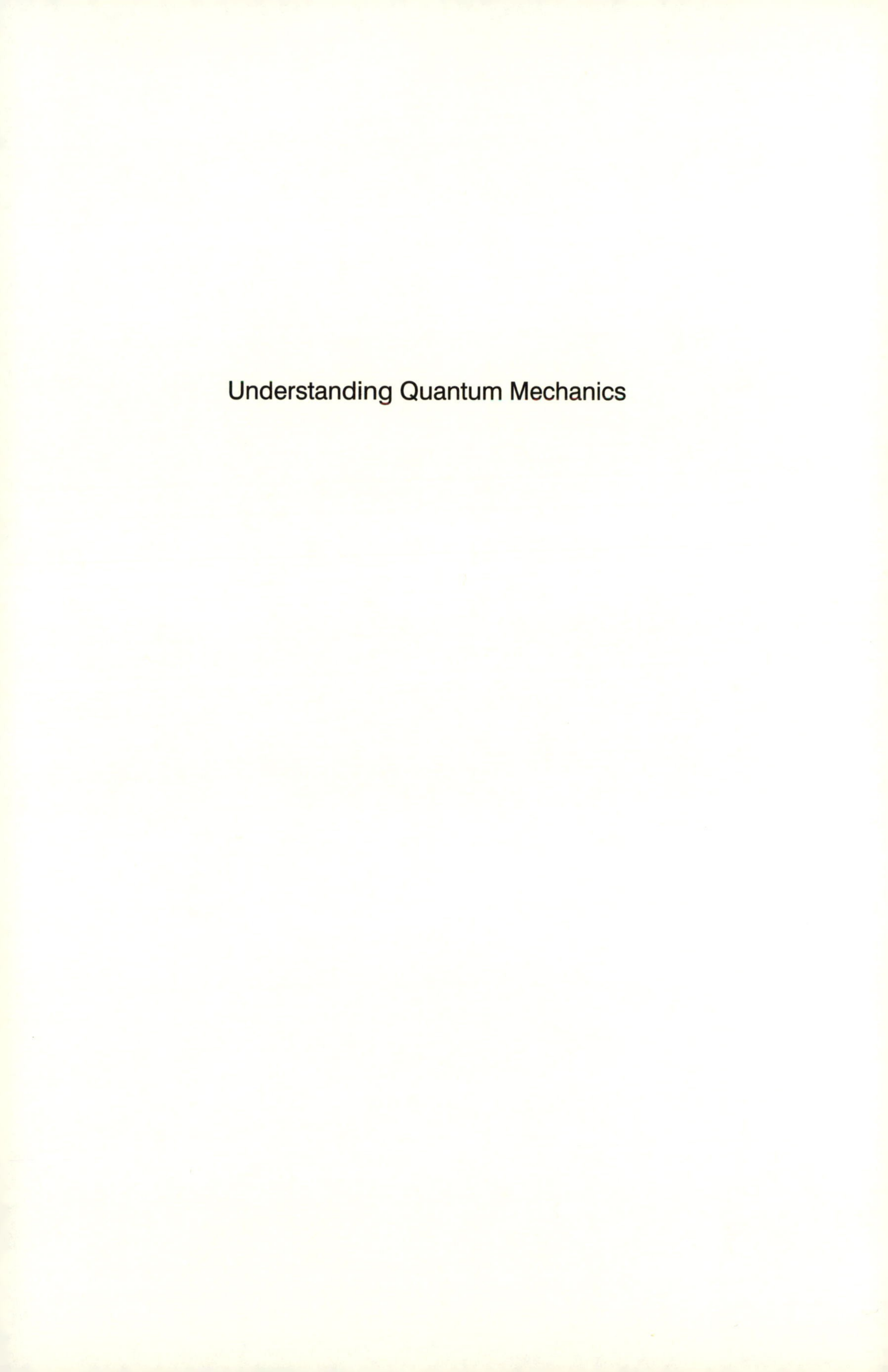

Understanding Quantum Mechanics

Understanding Quantum Mechanics

ROLAND OMNÈS

PRINCETON UNIVERSITY PRESS
PRINCETON, NEW JERSEY

Published by Princeton University Press, 41 William Street,
Princeton, New Jersey 08540
In the United Kingdom: Princeton University Press,
Chichester, West Sussex

ISBN 0-691-00435-8

Library of Congress Cataloging-in-Publication Data

Omnès, Roland.
Understanding quantum mechanics / Roland Omnès.
p. cm.
Includes bibliographical references and index.
ISBN 0-691-00435-8 (cl. : alk. paper)
1. Quantum theory. I. Title.
QC174.12.0465 1999
530.12—dc21 98-42442

This book has been composed in Times Roman with Helvetica Display

The paper used in this publication meets the minimum requirements of
ANSI/NISO Z39.48-1992 (R1997) (*Permanence of Paper*)
http://pup.princeton.edu
Printed in the United States of America
3 5 7 9 10 8 6 4

To Liliane

Contents

Preface

The aim of this book is to explain the interpretation of quantum mechanics, simply and completely if possible. It is addressed to beginners with some elementary knowledge of quantum theory and also to physicists wishing to refresh their memory and see what the present state of interpretation is.

There are many textbooks on quantum mechanics, but most of them do not give much space to interpretation and a book such as this one might be a useful complement to them. One of the meanings of "interpreting" is "understanding," and we all know that this is the main difficulty of quantum mechanics. Can it really be understood ? I think now it can.

Bohr, Heisenberg, and Pauli laid the foundations for interpretation long ago, and nothing that has since been discovered really undermines that groundwork. Some portions of their work have been revised, however, and others were found to be incomplete. Rather than replicate the writings of the founding fathers here, it would certainly be more faithful to their spirit by updating them rather than by repeating them.

Two important discoveries led to a revision of the old interpretation. The first one was the notion of decoherence, an effect explaining why no quantum interference can be seen at a macroscopic level and removing the perennial difficulty known as the Schrödinger cat problem. Decoherence was proposed in 1954 and was seen experimentally in 1996. It is now essential to an understand of quantum measurements. The second advance came in the last two decades from a full insertion of classical physics into the framework of quantum physics. Classical determinism was thereby found to be the direct consequence of a fundamental quantum probabilism, whose only remaining effect at our macroscopic level is that every causal prediction has a tiny probability of error. These results, the outcome of work by many people, shed a completely new light on interpretation.

A third important idea, consistent histories, was put forward in 1984 by Robert Griffiths and was not always clearly understood. Some people believed that it was an attempt (necessarily doomed to failure) for reintroducing a simple-minded realism into the quantum world. Histories have nothing to do with that. They rely directly on the basic principles of the theory and they give us a method, the means for clarifying and organizing interpretation, which can so easily turn into a quagmire, or a maze. Histories are also the foundations of a new language. One can do without them if one does not wish to change old habits, just as one can dig out the foundations of a building with a shovel rather than an excavator.

I wished to provide the reader with a convenient textbook and not a detailed treatise, and this aim entails a few consequences. I avoided difficult proofs and long calculations, replacing them as far as possible, by simple arguments. I refrained from recounting some trends of research, however deep or brilliant, that would be too speculative. The commentaries at the end of the book provide some recommendations for supplementary reading as well as for more technical information.

Physics will remain at the center of this book, but there will be little foray into epistemology. Philosophical issues cannot be completely ignored, however. I will not enter upon the difficult (or perhaps too trivial) question of realism because my emphasis will bear on the pragmatic aspects of physics. Even so, quantum mechanics requires a new way of understanding, as Bohr, Heisenberg, and Pauli knew, and if one is so bold as to take issue with Richard Feynman's famous saying—"nobody understands quantum mechanics!" (1965c, p. 129)—one must say not only how quantum mechanics can be understood, but also why, and what perspectives this understanding reveals.

I wrote another book on a similar topic a few years ago (*The Interpretation of Quantum Mechanics*, Princeton University Press), and I owe the reader an explanation for why a new one is proposed. First, there is a significant difference in emphasis: the present book is much less technical and more explanatory. A fair number of people kindly told me that something more accessible would be useful for students as well as for physicists with a cultural interest in interpretation and I tried to answer the request. A second reason is that research is active in this field and quite several significant results could be added.

Five important new features appear in this new book. The first one originates in a reading of the pioneers of interpretation and their commentators. I learned particularly from Dugald Murdoch and Catherine Chevalley of the inquiries by Bohr, Heisenberg, and Pauli on the language of interpretation. I realized better on this occasion the true nature of consistent histories and the existence of a wider frame for them. They are effectively a language for interpretation. This language has two faces, two versions: a direct one that can be expressed in plain English (or any other vernacular language) and an abstract one relying on mathematics and rooted in the basic principles of quantum mechanics. The second is a universal language: it can deal with atoms, particles, and also with experimental devices as well as with any ordinary object we can see or touch. It is also a rational language with a clear and explicit logical framework. When seen in that light, there is no possible confusion concerning the nature of histories. They are part of the universal language of interpretation. No language defines reality: it can only say what we know of reality. And we know much.

The second feature involves recent progress concerning decoherence. It has been observed, and more powerful theories now exist for it. Decoherence now appears to be an irreversible process, the strongest one in existence, and thus sheds a new light on time, logic, and irreversibility.

The third feature arose from a criticism of decoherence by John Bell and Bernard d'Espagnat. They said that decoherence cannot provide an explanation of the basic problems of interpretation but rather that it hides them by offering an answer valid "for all practical purposes," that is, an answer agreeing with any possible experiment and thereby forbidding us to see more deeply beyond these experiments. I came to the conclusion that these so-called fundamental questions hinge on the meaning of extremely small theoretical probabilities. An interpretation of probability calculus must therefore stand at the entry to an interpretation of quantum mechanics, and the most convenient one was proposed by Émile Borel: an event with too small a probability should be considered as never occurring. From an empirical standpoint, a very small probability is one which cannot be measured by any experimental device that can be realized, or even conceived, in the universe. One can then assert that decoherence cannot be bypassed and is really fundamental.

The fourth feature is also essential. When I wrote *the Interpretation of Quantum Mechanics*, the reader was left with a problem having no solution. This is the so-called *objectification problem*, which concerns the issue of the existence of a unique datum at the end of a quantum measurement. This looked like a big deficiency because many specialists claim that here lies "the" problem of interpretation. When looking at it, thinking of it, dreaming of it, I realized that it might be a false problem. It originates in a traditional line of thought that has no sure logical foundations. When decoherence is taken into account, one finds that the universal language of interpretation cannot encompass anything other than a unique datum, in agreement, of course, with observation. Said otherwise, there is no sound language in which the objectification problem might be stated, so therefore it may very well not exist. I must note that these ideas are an elaboration of previous insights by Robert Griffiths, Murray Gell-Mann, and James Hartle.

The last feature has to do with the meaning of probabilities and the randomness of quantum events. Probabilities enter the language of interpretation by the back door: they are found necessary for providing this language with a logical backbone. The relation with empirical randomness looks rather far-fetched, and it was not clearly explained in my previous book (as a matter of fact, it was not yet clearly understood, either). Here again, one finds that the language we must use cannot deal with anything other than random events. This may be not an explanation, like one could also say of objectification, but it is a logical necessity.

Finally, I may add that no effort has been spared for keeping the present account of interpretation as close as possible to the Copenhagen interpretation. There are four good reasons for doing so. As a complement of standard textbooks, this book would have been useless if it contradicted what one usually learns. It does not. Furthermore, the Copenhagen rules are used everyday in the practice of physics, and because they are correct, my task was to explain why without undue rewriting. It turns out also that histories can be considered as a systematic use of imaginary measurements, as used in Heisenberg's works. As to the last reason, I thought that showing another instance of two approaches, leading to the same conclusions through very different ways, is a beautiful example of the unity of physics.

To make the reader find his or her way more easily through this book, I may say how it is organized. There are three parts. The first one is a brief sketch of the early history of quantum mechanics. I am aware of a fashion in modern books to proceed as if everything worth knowing in science was discovered in the last decade, but it would not work here. On the contrary, I decided to start with a history, because some questions are much better appreciated when one sees them developing from the beginning, as well as how deep they go. There is a strong continuity in quantum mechanics and its interpretation, and I felt that the historical dimension is not only useful but also illuminating. Even if not answering them completely, one cannot ignore such questions as: How came this splendid theory, which needs another theory for being understood? Could it be otherwise? On the other hand, the Einsteinian question, "Should not it be otherwise?" is answered in the two following parts, showing the consistency of interpretation and the completeness of standard quantum mechanics.

The second part of the book is another brief history, this one of interpretation. Interpretation is often confusing and, I tried to provide an orientation.

There are three chapters in this second part. The first one is devoted to the Copenhagen interpretation, a name I propose reserving for the ideas expressed by Bohr, Heisenberg, and Pauli—and of nobody else. The second chapter is a summary of basic contributions, mainly by John von Neumann, that were not integrated in the standard Copenhagen interpretation. The most relevant criticisms by Einstein and Schrödinger are also discussed here. They had an essential role in keeping the question of interpretation alive and pointing out the most important questions. Other proposals and events, particularly around Bell's contributions, are also mentioned, even if very briefly.

The third chapter (Chapter 7 of the book) might be called a keystone, and the reader in a hurry might directly start from it. It is a nontechnical summary of the main aspects of a consistent interpretation of quantum theory establishing its completeness. It cannot be easily condensed, and I therefore will not describe it in this foreword.

The third part of the book is the longest one. It further develops the points indicated at the end of the second part. It does not refer to historical matters and gives a rather detailed and systematic account of interpretation.

The present book relies heavily on essential works, particularly by Murray Gell-Mann, Robert Griffiths, Serge Haroche, James Hartle, Jean-Michel Raimond, Hans Dieter Zeh, and Wojciech Zurek. It also relies on important ideas or advice from Roger Balian, Bernard d'Espagnat, and Lars Hörmander. I benefited immensely from exchanges with these wonderful people. Many others—too many to thank by name—helped me in this enterprise throughout the years.

I also wish to thank the staff of Princeton University Press and Gail Schmitt for her treatment of the manuscript, particularly her improvement of my English.

PART ONE

The Genesis of Quantum Mechanics

1

The Discovery of Planck's Constant

WHY ONE RELIES ON HISTORY

1. There is so wide a chasm between classical and quantum physics that one still marvels that it could be bridged. These two conceptions of the world are almost opposite in every respect. Classical physics relies directly on a reality that one can see and touch, on which one can act. It is basically causal, to the extreme of determinism. On the other hand, quantum physics deals with a world inaccessible to our senses and our common sense, a world that can only be described by abstract mathematics. The most commonplace objects have a shadow of wave functions, simple quantities become matrices, operators, pure constructs of pure mathematics. In place of the excessive certainty of determinism, one is facing absolute randomness. Could there be a more drastic opposition?

These two visions of the world agree, however, and it is now known that classical physics is a consequence of quantum physics. Under everyday circumstances, in our familiar surroundings so far from the atomic scale, quanta become classical. The question of how that happens cannot be separated from a preliminary one, which is how humankind succeeded in bringing to light an atomic world so far from intuition; that is, how its strange concepts were discovered. Our main task in this book will be to understand those concepts, and to do so requires that we look at their origin and make sure of their peremptory necessity.

During the nineteenth century, most chemists had become convinced of the existence of atoms, bound together in molecules and being transferred from one molecule to another during a chemical reaction. Physicists had also understood that the smallness of molecules can explain the laws of thermodynamics. There remained many nagging questions nevertheless. The existence of solid bodies showed that atoms cannot penetrate each other but that they can bind together in a molecule. What kind of force can be responsible for both attraction and repulsion? How can electrons find their way among the compact stack of atoms in a copper wire when there is an electric current? The list of questions was long, so long as to make Nietzsche speak of the existence of atoms as "the most refuted theory," but all the questions boiled down to a unique one: What are the laws of physics in the case of atoms?

The detailed story of the answer, the discovery of quanta, is complicated. It is full of distracting events, with ideas of genius intermingled with errors, and with many forbidding obstacles all along the way. This story holds the key for understanding why a very drastic change had to occur in science. If I were to leave history aside, I could perhaps describe more concisely the present state of knowledge, but with the desire for full understanding, we always need to retrace the old paths and again ask the old questions. The real difficulty in understanding an idea, a vision, or a concept can always be measured by the time it took for humankind to create and master it. This is why there is nothing better than the papers by the pioneers, because nobody else was more oppressed by the anguish of understanding. This is why perhaps understanding with no help from history requires more exertion and cleverness than having the patience to learn from it.

When history is used for the sole purpose of understanding science better, it must not get lost among details and must concern only what is really essential. This is why, while using the rigorous work of historians, I will look at a history of quantum mechanics that has been simplified with the help of later knowledge. This procedure would be a sin for a historian, who must reject anachronisms, but it may be justified when timeless laws (those of nature) are at stake and their existence can shed light on their discovery.

Perhaps the best summary of the story is Einstein's famous sentence: "The Lord is subtle, but He is not wicked." It looks as if there had been privileged pathways offering entries in the maze of atoms, a few fortunately simple and revealing problems, perhaps the only ones that could lead from the clarity of classical thought to the depth of quanta. History took these paths, and even now with our accumulated knowledge, we cannot conceive of other ones that could have been pursued. There were two providential problems: the harmonic oscillator and the hydrogen atom. Planck's constant, the touchstone of quanta, is revealed in its purity by the oscillator. The hydrogen atom is the other "paradigm," whose peculiar virtue is that it shows no essential difference between the deepest energy levels and almost classical ones. These two problems became the keys to quantum physics.

2. The first period in the history of quantum mechanics concerns the problem of the oscillator, which began in the middle of the nineteenth century, reached an apex in 1900, and was practically solved by 1911.

The answer is clear: some subtle laws of physics govern harmonic oscillators. They depend upon only one quantity, Planck's constant h, which is dimensionally an "action" (ML^2T^{-1}). It is extremely small ($6{,}6262 \times 10^{-34}$ joule/second). The only possible values for the oscillator energy are discrete and are given by $n\,h\nu$, n being an integer.

A few comments might be useful before describing how these results were obtained. It was surprising to find action playing a fundamental role in microscopic physics. No experimental device can measure an action. This notion appeared for the first time in mathematical theories with the principle of least action of Lagrange and Hamilton, which had obscure foundations. An oscillator is, on the contrary, a very simple system. It depends on a position variable x, with a kinetic energy $\frac{1}{2}m$ $(dx/dt)^2$ and a potential energy $\frac{1}{2}m\omega^2x^2$ so that x is an oscillating function of time $x_o \cos(\omega t\text{-}constant)$ with a frequency $\nu = \omega/(2\pi)$. The coordinate x is not necessarily a length, so that the inertia coefficient m is not always a mass (one may think of x as an electric charge; m is then a self-inductance). This freedom on the dimension of m may partly explain why the properties of a quantum oscillator do not depend much on this parameter.

Three different kinds of oscillators had a role in the story. The first type was abstract: a purely theoretical construct that was used by Max Planck as a tractable model of matter. The second included electromagnetic oscillators, also theoretical notions, which provided a convenient representation of radiation. Their properties were essentially equivalent to Maxwell's equations from the standpoint of dynamics and, when quantized, they became *photons*. The third species of oscillator was the only physically manifest one because it consisted of the elastic vibrations of a crystal. They became *phonons* in their quantum version.

BLACK BODY RADIATION

3. The first chapter of the story began in 1859, when Gustav Kirchhoff tried to explain the thermal radiation coming from the sun, a star, or a fireplace. The only theory at his disposal was thermodynamics, but he was able to apply it to this problem. A body in thermal equilibrium at a temperature T emits radiation through its surface. By $E(\nu)\Delta\nu$ I will denote the energy emitted per unit area and unit time in a frequency range $\Delta\nu$. Kirchhoff took into account the absorption coefficient $a(\nu)$ of the body, which is defined as follows: when the surface of the body receives from outside a radiation with frequency ν and intensity I, it absorbs a fraction $a(\nu)I$ and reflects the fraction $(1 - a(\nu))I$.

Kirchhoff then considered the following situation. Two infinite plates made of different materials are facing each other and are kept at the same temperature T. There is radiation between the plates and it is also in thermal equilibrium. Using simple arguments, Kirchhoff showed that equilibrium requires the ratio $E(\nu)/a(\nu)$ to be the same for all materials, and it must therefore be a universal quantity. For a perfectly black substance (with an absorption $a(\nu)=1$ for any frequency), the emissivity $E(T,\nu)$ coincides with this universal datum, whence the name of *black-body radiation.*

Later, in 1894, Willy Wien took another important step, again using thermodynamics. He considered a vessel at a temperature T in which there is a perfectly reflecting sphere ($a(\nu)=0$). The sphere first expands, then contracts, slowly enough for the process to be adiabatic. There is a slight Doppler shift when thermal radiation is reflected on the moving surface of the sphere, and this is enough for producing some energy exchange between neighboring radiation frequencies. Using the reversibility of the complete cycle, Wien found that $E(T,\nu)$ depends on a unique variable. The result is formulated more precisely as Wien's displacement law

$$E(T,\nu)=\nu^3 F(\nu/T). \tag{1.1}$$

The function F appeared to be a fundamental feature of a natural law, and many experiments were conducted to obtain it while many theoretical investigations tried to predict it. I will not recount the difficulties that were encountered nor the necessary advances in measuring techniques that were made but will mention only an important result by Rayleigh, who apparently found the function F in June 1900. He used a standard result from classical statistical physics (the equality, or "equipartition" of energy among all the degrees of freedom) and applied it to the radiation oscillators. He found F to be a constant. The result is perfectly correct from the standpoint of classical physics but at the same time perfectly absurd because his expression predicts that the total radiation energy $\int E(T,\nu)\,d\nu$ is infinite. The prediction had, of course, little to do with experimental data, except at very low frequencies. Therefore, Rayleigh's assumptions, those of well-established classical physics, had to be questioned in this situation. The problem was still more important than had been anticipated.

4. For several years, Planck had been trying to obtain a theory of black-body radiation. He placed special emphasis on the fact that thermal radiation in a vessel does not depend upon which kind of material the vessel walls are made of. Why should one then worry about real matter, so complicated and so little known at that time? Why not use an ideal case? Planck therefore considered an abstract model

consisting of harmonic oscillators with various frequencies. (A model of matter made of oscillators might have been almost realistic. J. J. Thomson proposed soon after—in 1903—a model where each electron behaves like a three-dimensional oscillator.) He had at his disposal a preliminary result, which is clearer when expressed in a later version given by Henri Poincaré: the average energy of an oscillator in matter must be the same as the average energy of an electromagnetic oscillator having the same frequency. This result relied only on the laws of electrodynamics, and the problem of thermal radiation could then be reduced to a single question: what is the average energy U of a single oscillator (U being proportional to νF)?

In October 1900, Planck found an empirical formula that perfectly fitted the data:

$$U = \frac{h\nu}{\exp(h\nu/kT) - 1}, \tag{1.2}$$

where k is Boltzmann's constant. Planck had obtained this formula by trial and error, and one can understand his approach by noting that the function (1.2) is very simple when the ratio $h\nu/kT$ is either large or small. Planck knew how to obtain these limiting expressions by assuming the oscillator entropy S to be proportional to either U or U^2. *Interpolating* between these two expressions of entropy (i.e., setting $S = U^2/(a + bU)$), he had only to solve an elementary differential equation for obtaining U (since $dU = TdS$) to get equation (1.2). The essential novelty of the result is the occurrence of a new constant h, such that the product $h\nu$ is an energy, just like the quantity kT. This empirical approach gave no hint, however, about the meaning of the formula. It said that an energy $h\nu$ is important, but nothing in the known theory of a classical oscillator had anything to do with it. Nevertheless, it must have been the key feature missing in Rayleigh's argument.

Planck then had a stroke of real genius. Could it be that the oscillator energy is not a continuous quantity that changes gradually, as required by classical physics? Could it take only discrete values, integral multiples of $\varepsilon = h\nu$? If so, the calculation is easy. Boltzmann showed that the probability for a system at equilibrium to have an energy E is proportional to $\exp(-E/kT)$. If E can take only the values $nh\nu$ with n an integer, the corresponding probability p_n is proportional to $\exp(-nh\nu/kT)$, and the average energy of the oscillator is given by

$$U = \sum nh\nu p_n \Big/ \left(\sum p_n\right). \tag{1.3}$$

The result is in perfect agreement with equation (1.2) and was published by Planck in December 1900.

As a matter of fact, the calculation I have sketched is not exactly the one Planck made, but his result was immediately reconsidered in many different ways, among which was the one I mentioned. Soon after, Poincaré proved mathematically that the distribution (1.2) cannot be obtained if one does not assume that energy is "quantized."

PHOTONS

5. Planck believed that "matter" oscillators were quantized but that radiation oscillators were not. One may easily understand his standpoint. Soon after Planck's discovery, the most favored models of atoms considered them as a collection of oscillators. So little was known about atoms that they could bear the burden of explanation, whereas on the contrary, radiation was perfectly described by Maxwell's equations and it looked unquestionable. This solution did not convince Einstein. The equal average values for the oscillators in matter and in radiation suggested a bold premise: why not assume that the radiation oscillators are quantized?

This idea was supported by the properties of the photoelectric effect: When visible light or ultraviolet radiation strikes a metal plate, some electrons are extracted from the metal and are seen as an electric current. The number of electrons extracted per unit time is proportional to the radiation intensity, a property that can be attributed to energy conservation. Much more surprising is the all-or-nothing effect that occurs with a change in the radiation frequency. If the frequency is lower than some value ν_o (depending on the material), no current is produced. Above that value, the effect is fully active.

Einstein proposed an explanation for these observations in 1905. If one assumes that in a radiation with frequency ν the energy is carried in "grains" with an energy $h\nu$, then everything becomes clear. From the known electron emissivity of a hot metal plate, one knows that each electron needs a minimal energy W for getting out of the metal. A grain of light must therefore carry an energy larger than W for extracting an electron, wherefrom the threshold effect when W is identified with $h\nu_o$.

The existence of light "grains" or light "quanta," later to be known as *photons*, was much more difficult to accept than Planck's quanta. The community where Einstein's idea was best received, though not fully accepted, was among specialists of X-rays, because hard X-rays can leave straight-line tracks in a photographic emulsion and this is strongly reminiscent of a particle trajectory. The photon hypothesis was not fully supported by experiments until 1923, when Arthur Compton found that X-rays diffracted by electrons show a frequency shift corresponding exactly to a relativistic recoil.

THE SPECIFIC HEAT OF SOLID BODIES

6. The concept of quanta of matter seemed strange, but this could be attributed to the mystery of atoms. Perhaps, after all, they were only accounting for appearances, which might receive a better explanation with further research. As for light quanta, they were almost unbelievable. Yet, by the time of the Solvay Meeting in 1911, everybody, including the foremost physicists of the day, accepted the existence of quanta. One reason for this almost unanimous adoption is that it had been shown there existed real, undeniable oscillators clearly demonstrating energy quantization: the elastic vibrations in a solid body.

X-ray diffraction had shown that the atomic structure of crystals are regular lattices of atoms (or molecules). Every atom has a definite position in the lattice that minimizes the energy of the structure. If an atom is slightly drawn away from its equilibrium position by a small distance a, the change in potential energy is proportional to a^2 (because energy would not be minimal if it contained a term linear in a). It can be shown in these conditions that small collective motions of the atoms are elastic waves with a definite frequency, each wave being mechanically equivalent to an oscillator.

When applying Planck's formula to these oscillators, one easily obtains the average energy of a solid body at thermal equilibrium and, as a consequence, its specific heat. This was shown for the first time by Einstein in 1907, who used a rather crude model in which all the vibrations are given the same frequency. Some time later, Peter Debye improved the model by using a more reliable distribution of frequencies. The results were clear and convincing, and they explained the vanishing of specific heats at low temperatures, which had remained an unsolved puzzle that contradicted classical statistical mechanics.

In a later development in 1916 by Einstein and again concerning thermal radiation, he assumed that the photons can be absorbed *randomly* by an atom. The probability for absorbing a photon in a given mode (i.e., with definite wave-number and polarization), is obviously proportional to the number of photons N in that mode. As for emission, Einstein assumed it to be a kind of decay of an excited state, and he applied the empirical law for the radioactive decay of nuclei, thus obtaining a distribution in time for emission. In order to fall back on Planck's law, one must assume furthermore that the probability of emission is proportional to $(N+1)$. This surprising result can only be understood if one assumes that the $(N+1)$ photons existing after emission are perfectly identical, with no memory of which one of them has just been emitted. Probabilities thereby entered the physics of quanta for the first time, but their subtleties remained hidden behind the poorly understood law of radioactive decay. The idea of indistinguishable particles would remain anyway.

7. It is sometimes said that the early history of quantum physics is obscured by thermodynamics, yet nowadays these questions are approached in the opposite order. Thermodynamics follows from statistical mechanics, which deals with a large number of particles obeying quantum laws.

Harmonic oscillators were central to the first period in the history of quanta. They are so simple that no contradiction of classical physics could have been as clear, and they still remain by now the simplest and most versatile concept in quantum physics. This is because they involve only Planck's constant, and the quantization of energy is as simple as it could be: only integers enter into the calculations. Once it was elaborated, quantum mechanics found that the exact formula for energy is $(n+\frac{1}{2})h\nu$ rather than $nh\nu$, but the correction is of little ado. Discrete energies exist, and their complete disagreement with classical conceptions means that the latter are either incomplete or erroneous. Physics would have to explore the first alternative before being compelled to accept the second.

2

The Bohr Atom

8. The second period in the history of quanta began in 1913. It ended in 1925 when a new theory of quanta, our present quantum mechanics, was taking over. This is the time when the path of physics research is most obscure. The pioneers entered the quantum world with classical minds; they maintained almost all of the classical laws of physics, adding only a few special ones for atoms. It is easy for us now to see how delusive this program was and also how fortunate its final outcome was. It made the failure of classical thought obvious and brought the real difficulties to the surface. Still, it was only a prelude to the advent of a radically new mode of thought.

The overall results at the end of this period were quite valuable. Essential experimental data were obtained, and the mixture of success and failure in explaining them brought imagination and subtlety to the forefront. It is not easy, however, to recount the course of events in a tumultuous period during which ideas came and went. The essential point is perhaps that everything was tried by some of the best people in the entire history of physics to save the classical vision of the world, and yet they failed. Trying to be complete or precise in my narration would be practically impossible given such conditions. I will therefore mention only the main results that survived and try to convey how the search was successful.

THE BOHR MODEL FOR HYDROGEN

The Atom's Nucleus and Rutherford's Model

9. What does an atom look like? For a short time, a model constructed by J. J. Thomson in 1903 was favored. The atom was supposed to be a positively charged, homogeneous sphere inside which electrons moved. The most attractive feature of this model was that an electron behaves in these conditions like a three-dimensional harmonic oscillator and the corresponding frequencies were tentatively identified with the spectral ones.

A quite different structure of atoms resulted from some experiments with radioactivity, that were carried out by Hans Geiger and Ernest Marsden in Thomson's laboratory. In 1909 they observed a remarkable effect: when alpha particles (coming from a radioactive source) cross a thin metal plate, their trajectories can show strong deviations. In 1911 Ernest Rutherford determined the significance of this effect. A strong deviation of the trajectory meant that "something" inside the atom exerts a strong force on the alpha particle, and also that this something is heavy enough for sustaining the recoil. Electrons are too light for that. If the force is electric, it must be exerted at very short distances to be strong enough, so the scattering object must be very small, charged, and heavy, just like an alpha particle. This idea of something at the center of atoms—a nucleus—was checked by Rutherford. After a rather easy calculation of the various deviation probabilities, the agreement between the model and observation was excellent.

The analogy between Coulomb forces and gravitation forces implied strong similarities between an atom and a minute solar system. Nothing could have been simpler, at least at first sight, but hard problems were lurking ahead.

Atomic Spectra

10. Atomic spectroscopy had been the basis of Thomson's model. On the contrary, the one Rutherford proposed was very far from explaining anything in that field. Spectroscopy was nevertheless a well-developed science. The frequencies of emission (and absorption) lines in most atoms were known, at least in the visible domain and its immediate vicinity. Some relationships had also been seen among the data. In 1890 Johannes Rydberg made a discovery that became a principle of spectroscopy: The emission (and absorption) frequencies of an atom can be labelled by two numbers, n and m, and a frequency ν_{nm} can be written as a difference, $\nu_n - \nu_m$, between two reference frequencies in a catalogue $\{\nu_n\}$. In the case of hydrogen, a simple formula had been found by Johann Balmer (and also probably by Rydberg): $\nu_n = R_o/n^2$, n being an integer and R_o the Rydberg constant. But no one could provide an interpretation of these perplexing results.

Bohr's Model

11. The Rutherford model is particularly simple in the case of a hydrogen atom, having a single electron in Keplerian motion. Niels Bohr, a young Dane working in Rutherford's laboratory, however, made

a troubling observation: the electron is accelerated during its motion. According to Maxwell's laws of electromagnetism, an accelerated charged particle should radiate and thereby lose energy. The binding of the electron must increase to compensate for that loss, and as a result the electron will come closer and closer to the nucleus, finally falling onto it. A simple calculation shows that the collapse of the atom must be extremely rapid, almost instantaneous and, moreover, the resulting continuous radiation would have no similarity with the observed line spectrum.

This was not, fortunately, the first instance of a breakdown of classical laws. Planck and Einstein had already found ways out in similar circumstances. Rather than giving up Rutherford's model, Bohr wanted to take into account the existence of Planck's constant. The oscillator example, with its quantized energy levels, suggested that atom energies could also be quantized. Letting E_n denote the energy levels of hydrogen and using an analogy with the photoelectric effect, Bohr assumed that a photon can only be emitted when the atom energy falls from a level E_m to a lower one E_n. Energy conservation requires that the photon energy h_ν be equal to $E_m - E_n$, and one then gets immediately Rydberg's rule:

$$h\nu = (E_m - E_n); \quad \text{with } \nu = \nu_{mn} = \nu_m - \nu_n = E_m/h - E_n/h \quad (2.1)$$

Because of the reversibility of the basic laws of physics under time reversal, emission is the process inverse of radiation, and the emission lines and absorption lines coincide. There is, therefore, no continuous spectrum, and the atom is stable in its lowest energy state, assuming there is one.

The energy levels E_n were known from the data (up to an additive constant), but the problem was understanding their values. In a classic article published in 1913, Bohr derived them in three different ways. They all yielded the same result, although none of them can be said to be really compelling. In a footnote Bohr mentioned that his results might also be obtained if one assumes that the line integral $\int p\,dl$ along the trajectory (p being the electron momentum and dl an element of length) is a multiple of Planck's constant:

$$\int p\,dl = nh. \quad (2.2)$$

The result one obtains for the energy levels, using only circular trajectories, is

$$E_n = \frac{me^4}{2\hbar^2 n^2}, \quad (2.3)$$

where m is the electron mass and e its charge. In the MKSA system of units, which is rather inconvenient for atomic physics, e^2 should be replaced by $e^2/(4\pi\varepsilon_o)$. The modern notation $\hbar = h/(2\pi)$ is also used in this equation.

First Confirmations

12. Bohr's results were in satisfactory agreement with the hydrogen data. The derivation of Rydberg's constant was enough for Einstein to recognize immediately the importance of the new discovery. Two startling events would soon convince most other physicists. First there was the solution of the ζ Puppis enigma. The star ζ Puppis had shown lines very similar to the hydrogen spectrum, except that R_o had to be replaced by $4R_o$. It was easy for Bohr to replace the hydrogen nucleus charge e by $2e$, the charge of a helium nucleus (the factor e^4 in equation (2.3) being replaced by $(Ze)^2e^2$ with $Z = 2$). The spectral lines of ionized helium had not yet been observed under laboratory conditions, but as soon as they were, Bohr's explanation was a complete success.

The other event, which convinced many sceptics, came from highly precise spectroscopic data. Equations (2.2) and (2.3) did not quite agree with the hydrogen data, which contained a slight error of the order of 1/2000. In classical mechanics, however, an electron does not revolve around the center of the nucleus, but around their common center of mass. Inertia is not governed by the electron mass m but by an effective mass $m' = mM/(m + M)$, where M is the nucleus mass. The correction is of the order of m/M, which is 1/1836 for hydrogen. When Bohr made the correction, the agreement with the data became almost perfect.

THE BOHR-SOMMERFELD THEORY

13. The conceptual framework of Bohr's theory had as yet little to do with quantum mechanics as we know it. Classical concepts were still directing the representation of atoms. They involved the electron trajectories, the motion itself, and the expressions giving the kinetic and potential energies. Planck's constant entered only in a supplementary condition, which selected the few permissible trajectories among the many possible ones. One can therefore characterize this kind of physics as classical physics with quantum addenda.

Bohr himself was far from convinced that his solution was akin to a final answer, and he believed that a deeper revision of physics would have to be made. Before doing so, however, the main task was to increase and organize the collection of experimental data, as was done from 1913 to 1925.

The Correspondence Principle

14. In his first paper Bohr had put forward a principle that was to become more and more important. It was the *correspondence principle*, which states that the laws of quantum physics must reduce to those of classical physics when quantum numbers, such as n, become large. As an example, one may consider one of the derivations Bohr had given for equations (2.2) and (2.3), using this principle (it was, by the way, the most convincing one). A charged particle having a classical periodic motion with frequency ν can only emit radiation whose frequencies are multiples of ν. This result follows from the emission theory of Maxwell and Hertz and, more generally, from a Fourier analysis of the motion. It is far from being satisfied by Bohr's model, but when the electron trajectory is big enough to look classical (or n large enough), one can apply the correspondence principle in the following way: The nearest one can approach classical continuous motion is to have the electron jump from a trajectory with a large value of n to the next one with the quantum number $n' = n + 1$. There is emission of a photon during the jump, and according to the principle, the photon frequency should be almost exactly the frequency of classical motion along trajectory number n. This gives equations (2.1–3), although an important proviso must be made: the "proof" is valid for large values of n, but the result is used for any value of it. What luck that it could work!

Sommerfeld's Rules

15. To turn Bohr's discovery into a general theory, it had to be extended to every atom and, in the case of hydrogen, include elliptic trajectories because Bohr had only considered circular ones. Arnold Sommerfeld solved the second problem in 1915 and, for some time, there was hope that the first one, although much more difficult and more general, might also be solved along the same lines.

Sommerfeld considered a physical system and described it in a standard classical manner by general time-dependent coordinates $q_k(t)$. They are not necessarily Cartesian coordinates with the dimension of a length; they can be, for instance, angles (if one thinks of a top, or rigid rotator, in which case Euler angles are the coordinates). Because Planck's constant has the dimension of an action, momenta in equation (2.2) must be replaced by something else having the correct dimension.

What else? Being well aware of classical mechanics, Sommerfeld introduced the Lagrange and Hamilton's conjugates of the coordinates.

Let us recall the principle: The position of a physical system is defined by some variables $q=(q_1,q_2,\ldots,q_N)$, velocities being the time derivatives $\dot{q}=\{\dot{q}_1,\dot{q}_2,\ldots,\dot{q}_N\}$. Energy consists of a kinetic energy $T(q,\dot{q})$ (depending generally on q and $\dot{q}$), and a potential energy $V(q)$. The moment conjugate to the position variable q_k is defined by

$$p_k=\frac{\partial T}{\partial \dot{q}_k}. \tag{2.4}$$

It has automatically the right dimension for the product $p_k q_k$ to have the dimension of an action. One can draw from these equations the velocities $\dot{q}$ in terms of the moments $p=(p_1,p_2,\ldots,p_N)$ and the coordinates q. When written as a function $H(q,p)$ of these variables, the total energy $T+V$ is called the *Hamilton function* and the equations for dynamics become the *Hamilton equations*:

$$\frac{dq_k}{dt}=\frac{\partial H}{\partial p_k}; \quad \frac{dp_k}{dt}=-\frac{\partial H}{\partial q_k}. \tag{2.5}$$

This version of classical mechanics, more abstract than Newton's, became thereby more familiar to physicists and would soon become the language of quantum mechanics.

In the case of the hydrogen atom, the most natural position coordinates are the Cartesian coordinates of the electron, the conjugate moments being the components p_k of ordinary momentum. One can also use spherical coordinates (r,θ,ϕ). On an elliptic trajectory, the azimutal angle ϕ varies from 0 to 2π, the angle θ and the radius r oscillate back and forth between some extreme values. Sommerfeld's most important contribution was to notice that one can replace the unique condition (2.2) by a system of three conditions

$$\int p_r\, dr=n_1 h, \quad \int p_\phi\, d\phi=n_2 h, \quad \int p_\theta\, d\theta=n_3 h, \tag{2.6}$$

in which the integration over r and θ are performed on one oscillation (back and forth) of the corresponding variable whereas the integration over ϕ goes from 0 to 2π. The quantities n_j are integers.

These conditions can be combined so as to return to Bohr's condition (2.2) with $n=n_1+n_2=n_3$. The other two quantum numbers have a simple interpretation: The number n_3 (usually denoted by m and not to be confused with a mass) is related to the z-component of angular momentum by $L_z=m\hbar$. The quantity $(n_2+n_3)^2\hbar^2$ is the square of the angular momentum vector $L^2=l^2\hbar^2$. The three quantum numbers (n,l,m), characterize an energy state of hydrogen, and they soon became familiar to everybody.

By introducing elliptical Keplerian trajectories, Sommerfeld had made significant progress. He went further by considering a more general system with coordinates $\{q_k\}$, under the assumption that the kind of oscillatory motion for each coordinate, which is true for hydrogen, is still valid. The quantization rules become in that case

$$\int p_k \, dq_k = n_k h. \tag{2.7}$$

Unfortunately, the assumption of a multiperiodic motion is extremely restrictive and the physical systems for which it holds are quite exceptional. For instance, they do not include an atom with several electrons.[1] They do include, however, most systems for which the motion can be computed explicitly by analytical methods. These systems were the best known ones at that time and gave a temporary illusion that a generalization would be found. As time went on, however, this hope faded until it vanished completely, giving way to the hard truth: Sommerfeld's conditions required assumptions that were too special to be fundamental laws of physics.

LANDÉ'S RULES

16. Two cases in which Sommerfeld's method was successful were its explanations of the Stark and Zeeman effects in hydrogen. The Stark effect is observed when an atom is subjected to an external electric field: spectral lines split and mix. The Zeeman effect occurs in a magnetic field and gives only splitting: a multiplet of spectral lines replace a unique line.

That was the time when Hamilton's formalism took precedence over the Newtonian form of classical dynamics. The choice of variables that allows one to quantize a system under Sommerfeld's rules is rarely given by the ordinary Cartesian coordinates for position and velocity. The Stark effect in hydrogen, for instance, requires parabolic coordinates.

[1] A precise definition of multiperiodic, or separable, systems relies on the Hamilton-Jacobi equation. It will be more in the spirit of this book to define them by their modern quantum version: they have a basis of energy eigenfunctions $\psi(q)$, each of which can be written as a product $f_1(q_1)f_2(q_2)\ldots f_k(q_k)$.

One of Sommerfeld's successes was a calculation of relativistic corrections to the hydrogen levels, in which he used a canonical change of coordinates and momenta, a move that is typical of the full power of analytical dynamics. The result—the fine structure of levels—was most encouraging.

The systematic experimental study of the Zeeman effect in various atoms was particularly fruitful. Using spherical coordinates along the magnetic field B, one finds that the levels in a multiplet are given by the simple formula (obtained from a classical perturbation calculus)

$$E_{nlm}(B) = E_{nl} - m\omega\hbar, \tag{2.8}$$

ω being the frequency eB/m_e. Since the quantum number m can take every integer value from $-l$ to $+l$, one has a direct access to the value of l by counting the spectral lines in a multiplet. Similar considerations hold for hydrogen, although the degeneracy in l requires some more care.

Theorists thus had an opportunity to consider much more data, and they arrived at a rather puzzling situation. Some atomic spectra agreed with theory, in which case, they were referred to as a "normal" Zeeman effect. Other data were at variance with the same theory, producing what was called an "anomalous" Zeeman effect. Unfortunately, there were more anomalous cases than normal ones. How can one look happy when thinking of the anomalous Zeeman effect? Pauli once said.

A second phase in research led to variations on Sommerfeld's rules that took more and more liberties with them. A touch of imagination and subtlety, with not too much belief in the foundations, became the rule. Alfred Landé was among the best people at that game. He elucidated, or contributed to the elucidation of, quite a few empirical rules. They were inspiring, but puzzling as well, and they can be summarized as follows: The z-component of an angular momentum can always be written as $m\hbar$, but m can be an integer or half an integer. The square of an angular momentum has a value $j(j+1)\hbar^2$, where j is again an integer or half an integer. Every atomic level in a magnetic field is characterized by four quantum numbers and not three.

I will not try to report how these results were obtained. They required much subtlety with a very good knowledge of the data, together with either direct use of the accepted theory or implications suggested by it. The most important point is that the results cast even more doubt on the Bohr-Sommerfeld approach. They would also play an essential role some time later when a new theory of quanta, the Heisenberg-Schrödinger formulation, would be able to explain them.

FROM ATOMIC SHELLS TO SPIN

Atomic Shells

17. A careful examination of all atomic spectra led Bohr to recognize, in 1921, that there is a simple structure in atoms. He compared the changes in spectra when the atomic number Z increases unit by unit. He attributed three quantum numbers (n, l, m) to each electron and investigated the distribution of these numbers in the ground state of each atom. There is a degeneracy for different values of m. In hydrogen, $Z = 1$ and the electron has the quantum numbers $n = 1$, $l = 0$ $(l \leq n - 1)$. It is the same for helium, $Z = 2$, with the two electrons. In lithium, $Z = 3$, two electrons are in a "shell" looking like helium and the third one has the quantum numbers $n = 2$, $l = 0$. Beryllium has two electrons looking like the third one in lithium and then, from boron to argon, six electrons have the quantum numbers $n = 2$, $l = 1$. I'll stop here, but Bohr carried out an examination of the entire Mendeleyev table.

He met subtle effects, which he could master with the help of Landé's rules and the kind of intuitive art that was so characteristic of the time. He did not make any calculations (that was never his hobby), but he could explain many chemical and physical data. The shell structure of atoms was the result, confirming the strong hints of it that had been found in 1916 in an important experiment by James Franck and Gustav Hertz in which electrons collided with atoms.

The Exclusion Principle

18. Bohr's results defined a clearcut problem: why are the electrons distributed among different shells rather than all sharing the deepest one so as to minimize energy? In 1925, Wolfgang Pauli proposed an answer, or rather, a rule: the exclusion principle. Two electrons at most can share the same quantum numbers (n, l, m). There was no good reason for that restriction and, even now, when Pauli's rule has become part of a larger symmetry principle, the latter stands by itself, independent of the other principles of quantum mechanics. In any case, people were most worried at that time by a much simpler problem: Why two electrons and not just one? Had this situation anything to do with the fourth quantum number one found in some places and the strange half-integers in Zeeman effect?

Spin

19. Two young Dutch physicists, George Uhlenbeck and Samuel Goudsmit, answered the question in 1925. Theirs was a bold idea: an electron has a proper angular momentum, a spin, corresponding to the value $j = 1/2$, and a projection along any space axis that takes only the two values $m_s = \pm 1/2$. The exclusion principle accordingly became clearer: there can be only one electron in a state with given quantum numbers (n, l, m, m_s).

The idea was not easy to accept. There was an older preconception according to which the electron was an electrically charged sphere with a radius of the order of 10^{-15} m (for which the electrostatic energy is $m_e c^2$). If spin means a rotation of that sphere, the velocity at the sphere surface would have to be greater than the velocity of light. So accepting the idea of spin meant giving up the only model of the electron and, perhaps worse, the idea that angular momentum accompanies a rotational motion. As a matter of fact, Henryk Kramers, had also previously conceived the idea of electron spin first, but he was convinced to renounce it after hearing these contrary arguments from Pauli.

The idea was accepted, nevertheless, for the best possible reason: it explained many data and disagreed with none. But in 1925, physics was already changing profoundly.

AN OVERVIEW

20. What can be said, finally, of the second period in the history of quantum physics? Little of the work produced remains, as least as far as the basic laws of physics are concerned. The exclusion principle and the existence of spin are retained. As for the would-be principles—Bohr's model and Sommerfeld's rules—they turned out to be much too specific with no hope of a general formulation, and too many data contradicted them. The framework they proposed, a classical description of atoms to which quantization rules were added, was finally rendered untenable.

A paper by Bohr, Kramers, and John Slater, published in 1924, can be considered as marking the end of this part of the story. They made it clear that one could not combine real electron trajectories with an explicit description of radiation emission and absorption. Doing so would contradict energy-momentum conservation.

Yet the period was very fruitful when viewed as a time of maturation and preparation. It produced much experimental data, and ephemeral theory was useful for organizing them and generating empirical rules, including quantum numbers, the properties of angular momenta, and atomic shells. Many of these would become touchstones for the next theory.

The Bohr-Sommerfeld theory was a bridge over a chasm but one that fell down after it was crossed. One may wonder how it could hold up so long and reach so far. The reason is probably that with our present knowledge of quantum mechanics, we can look across the chasm from the other side and see the semiclassical situations in which Planck's constant appears relatively small. When computing energy levels, for instance, there are two, exactly two, realistic physical systems for which semiclassical formulas can be extended to deep energy levels with no appreciable changes. They are the harmonic oscillator and the hydrogen atom, the only two for which a classical vision could hold long enough.

3

Matrix Mechanics

21. The failure of the "first quantum theory" was becoming obvious in the early 1920s. Bohr stressed in his lectures the urgent need for a deeper and bolder theory, even if a break with traditional thinking was necessary. The "new quantum theory" was discovered soon after in two versions that were quite different and whose histories are somewhat intermingled. First there was the proposal by Louis de Broglie in 1924 for a wave mechanics that associated a wave with every particle. It took some time before it received a proper formulation by Erwin Schrödinger in 1926. Meanwhile, in 1925, Werner Heisenberg proposed a matrix mechanics, which had soon become a full-fledged theory in a "Drei-Männer Arbeit," a work by three people: Heisenberg Max Born, and Pascual Jordan.

Whoever teaches quantum mechanics nowadays knows how difficult it can be to explain Heisenberg's approach and how much simpler the one by de Broglie and Schrödinger may look. In spite of that, I will begin with Heisenberg, and not only for historical reasons. In the realm of quantum mechanics, simplicity is often more or less an illusion. Beginning with wave functions may look easy, but real difficulties and the need for a drastic change in our thinking necessarily follow. Schrödinger himself was to have this experience. Heisenberg's approach, in which the notion of a physical quantity is reconsidered from the start, shows more plainly that a revolution of thought was taking place. His approach gives its true dimension to my main topic, which is interpretation.

I will therefore review, perhaps too rapidly, the first definitive version of quantum mechanics. Neither will I provide much detail and sometimes, for the sake of clarity, some arguments in the original papers will be replaced by others which came soon after. Moreover, I will not distinguish between Heisenberg's original paper and its common elaboration with Born and Jordan.

HEISENBERG MATRICES

22. There are three main sources for Heisenberg's ideas: Bohr's correspondence principle (see Section 14), Heisenberg's own previous experience in physics, and his feeling about the nature of physics.

Heisenberg's previous experience in research came mainly from a work he had done as Kramer's collaborator. In the framework of Bohr-Sommerfeld theory, they investigated the dispersion of light (i.e., the dependence of a refraction index on frequency) from the interaction of an individual atom with light. Heisenberg's conviction about physics was, at least at that time, strict positivism: a correct theory should only make use of observable quantities. He had been inspired by some statements he heard from Bohr and also by what he believed to have been Einstein's attitude when formulating the theory of relativity. It does not matter whether Heisenberg later had to change his position (which Einstein refused to recognize as his own). The essential point is that a belief, even if erroneous, can be propitious in exceptionally creative circumstances, and this one was.

Heisenberg's Tables

23. What could Heisenberg consider as sufficiently well established that it could be used as a starting point? There was the existence of "stationary" states of an atom, each one of them with a definite energy E_n. When an atom emits radiation and goes from a state n to another m, energy conservation requires the radiation frequency to be $\nu_{nm} = (E_n - E_m)/h$.

Heisenberg took his first inspiration from the correspondence principle by looking first at the classical emission of radiation, which is a resonance effect: a moving charge whose motion has a definite frequency emits radiation with that frequency. This is basically a mathematical relation following from Fourier analysis, and Heisenberg applied it to quantum phenomena. Some physical quantities properly belonging to the atom must have the frequencies ν_{nm} existing in radiation. Because of the two indices (n, m), there must be as many such quantities as there are couples (E_n, E_m) of energy levels.

When an atom is subjected to a magnetic field, all the energy levels are distinct, whereas there is degeneracy otherwise. One may therefore expect the existence of as many relevant quantum quantities as there are different couples of stationary states, and not only different couples of energy levels. The index n labels these states, and the collection of resonating quantum quantities to be considered can be written as

$$X_{nm}(t) = X_{nm} \exp(-2\pi i \nu_{nm} t). \tag{3.1}$$

It is a table of complex numbers with two entries n and m. The exponential factor $\exp(-2\pi i \nu_{nm} t)$ indicates their resonating time dependence. The letter X, for the time being, stands for an unknown object.

One can then use again the correspondence principle. One may consider a system analogous to a hydrogen atom in which there is a moving charge. The classical interaction with an external electric field $E(t)$ is $-qE(t)x(t)$, where x is the position coordinate of the charge along the electric field. In a classical version of the hydrogen atom, the time-changing position coordinate $x(t)$ must replace the set $\{X_{nm}(t)\}$. One can therefore assume that any physical quantity $a(t)$ should be replaced at the quantum level by a two-entry table $\{A_{nm}(t)\}$. This is obviously a very strong assumption, but not a difficult one to use when it is accepted.

Heisenberg's philosophical preconception helped him take the next step. From his positivistic standpoint, the position of an electron in an atom could not be measured. He considered this to be impossible as a matter of principle, and not only because of the limitations in available measuring devices. Accordingly, the unobservable quantity $x(t)$ must be left out of the pattern. As for the new quantities X_{nm}, his previous experience with research turned out to be helpful. With Kramers, he had already met similar objects having a correspondence at the classical level with an atom electric dipole (i.e., $ex(t)$ in the case of hydrogen). He believed one could reach these $eX_{nm}(t)$ by some measurements, and he was right as far as their absolute value is concerned: one can obtain them from a measurement of a refraction index. Heisenberg believed, moreover, that the phase of these quantities could also be measured, but this was an error; although a fortunate one. Such an error cannot be considered a mistake. As it happens sometimes in the history of physics, this one was most fortunate.

Heisenberg's Algebra

24. It is certainly a very strange idea to introduce a table of complex numbers in place of so simple and intuitive a quantity as the position coordinate of an electron. Even worse, there are as many different tables as different atoms. One may therefore wonder what remains, in view of the correspondence principle, of the classical conception of x as a number. An example will be helpful. Let us assume that, in place of

the linear coupling of x with an electric field, one can realize a quadratic coupling proportional to x^2. This is not too far-fetched an assumption because the pressure exerted on an atom by its neighbors has an effect of that kind. Whatever it may be, let us take it for granted. A new table $\{X^2_{nm}(t)\}$ must then be in correspondence with the physical quantity $x^2(t)$.

It goes without saying that $x^2(t)$ is the square of $x(t)$. Dimensional arguments imply that the quantities $\{X^2_{nm}(t)\}$ should also be quadratic functions of the $\{X_{kl}(t)\}$. This means that one has a relation

$$X^2_{nm}(t) = \sum_{klk'l'} c_{nmklk'l'} X_{kl}(t) X_{k'l'}(t),$$

with a priori unknown coefficients $c_{nmklk'l'}$. Because a product $X_{kl}(t)X_{k'l'}(t)$ has the frequency $(E_k - E_l + E_{k'} - E_{l'})/h$, which must be the same as the frequency of $X^2_{nm}(t)$ if the product enters in the expansion, the previous equation can be reduced to

$$X^2_{nm}(t) = \sum_k c_{nmk} X_{nk}(t) X_{km}(t).$$

One can then make a virtue of the multiplicity of the quantum tables $\{X_{nm}\}$ for a unique classical quantity (with as many tables as there are atoms). The classical algebraic relation between x and x^2 does not depend on the physical system to which the electron with coordinate x belongs. If the same is true for the previous equation, it follows that the coefficients c_{nmk} are simply some algebraic numbers. Which ones? Heisenberg set them all equal to 1, saying that this was an "almost unavoidable" assumption according to the correspondence principle. He therefore finally arrived at

$$X^2_{nm}(t) = \sum_k X_{nk}(t) X_{km}(t). \qquad (3.2)$$

One can recognize in this relation the multiplication rule for two (infinite) matrices. Heisenberg had no inkling of matrices, but Born knew them and Jordan was an expert on them. The most puzzling aspect of the rule when extended to the product of two different quantities was, of course, noncommutativity.

DYNAMICS

25. Quite remarkably, noncommutativity turned out to be the key for quantum dynamics. The first step in that direction came from a consideration of the table representing energy: it is diagonal. This is strongly

suggested by the fact that an energy E_n depends only on one index n. One can look at it differently in the following manner: If the unit of energy is slightly changed, being multiplied by $1+\varepsilon$ with ε a small number, E_n becomes $(1-\varepsilon)\ E_n$. One could also imagine that the atom's energy E has been perturbed by adding to it $-\varepsilon E$. The perturbation does not change the stationary states because it has no physical reality. There are therefore no nondiagonal elements such as $-\varepsilon E_{nm}$ with $n \neq m$, which would perturb the stationary states. Said otherwise, the physical quantity "energy" to be now denoted by H (as in Hamilton) is represented by a diagonal matrix, whose elements $\{E_n\}$ do not depend on time.

The rules for dynamics follow immediately. One may ask, for instance, which table is in correspondence with the velocity component $v = dx/dt$. The simplest assumption is that the table $\{V_{nm}(t)\}$ is given by the time derivative of $\{X_{nm}(t)\}$, from which one gets

$$V_{nm}(t) = \frac{d}{dt}X_{nm}(t) = -2\pi i \nu_{nm} X_{nm}(t) = -\frac{2\pi i}{h}(E_m - E_n)X_{nm}(t)$$
$$= \frac{i}{\hbar}(E_n X_{nm}(t) - X_{nm}(t)E_m).$$

When written in matrix calculus, this relation becomes

$$\frac{dX}{dt} = \frac{i}{\hbar}(HX - XH) \equiv \frac{i}{\hbar}[H, X]. \tag{3.3}$$

This is easily extended to the time derivative of any physical quantity. Since the matrix H commutes with itself, one has automatically $dH/dt = 0$, that is, energy conservation. A conserved classical quantity must be similarly in correspondence with a matrix commuting with H.

CANONICAL COMMUTATION RELATIONS

26. The theory is still extremely formal, and nothing has yet been computed from it. How is one to obtain the energy levels, for instance? One may consider the simple case of a free electron in a one-dimensional space, with position and momentum represented by two matrices X and P. The correspondence principle suggests taking $H = P^2/2m$. From the dynamical rule (3.3), one gets

$$\frac{dX}{dt} = \frac{i}{2m\hbar}[P^2, X] = \frac{i}{2m\hbar}(P[P, X] + [P, X]P).$$

The classical relation $dX/dt = P/m$, suggested by the correspondence principle, is easily obtained if one has

$$[P, X] = \frac{\hbar}{i} I, \tag{3.4}$$

where I is the identity matrix. The wide validity of the *canonical commutation relation* (3.4) is then easily checked: Let one assume that the particle is classically in a potential $V(x)$. If this is a polynomial in x, or a series $a_o + a_1 x + a_2 x^2 + \cdots$, the algebraic rules of the theory turn it into a matrix $V(X) = a_o I + a_1 X + a_2 X^2 + \cdots$, which is well defined. From equation (3.4), one then obtains

$$[P, V(X)] = \frac{\hbar}{i} V'(X),$$

where $V'(x) = dV(x)/dx$. Using now for the energy the expression $H = P^2/2m + V(X)$, one gets

$$\frac{dX}{dt} = P/m; \quad \frac{dP}{dt} = -V'(X), \tag{3.5}$$

that is, the Hamilton equations now written in a matrix form.

For a system with several coordinates $\{Q_1, Q_2, \ldots\}$ and conjugate momenta $\{P_1, P_2, \ldots\}$, the canonical commutation relations are similarly given by

$$[P_k, Q_k] = \frac{\hbar}{i} I, \quad [P_j, Q_k] = 0 \qquad \text{for } j \neq k. \tag{3.6}$$

Predictions from the Theory

When completed by the canonical commutation relations, the theory can be used for a computation of the energy levels or the matrix elements of X and P, at least in principle. Such calculations may require some algebraic virtuosity, and they have now been replaced in most cases by more tractable methods using wave mechanics. One nevertheless can still find in modern textbooks some of the early calculations that were made by Born, Heisenberg, Jordan (and Pauli): for example, the computation of eigenvalues for the angular momentum quantities J^2 and J_z, together with the matrices representing J_x and J_y. The standard calculation for the energy levels of an oscillator is also familiar, and it is still required in quantum electrodynamics.

OUTLOOK

27. From matrix mechanics was born a new physics and almost everything coming afterwards is, in one way or another, a consequence of it.

The "new quantum theory" broke with traditional modes of thought in both obvious and subtle ways. It clearly rejected intuition and particularly the standard visual representation of a physical object. This was not too much of a surprise after the dismissal of electron trajectories following the work by Bohr, Kramers and Slater. More subtle were the surviving analogies between the new theory and older conceptions: for instance, when a photon is emitted or absorbed, the matrix elements show resonant transitions in a perfectly conventional way according to Fourier analysis and contrary to the Bohr-Sommerfeld model. Matrix products may also have an intuitive meaning: Similar products had already been encountered in the paper by Kramers and Heisenberg on dispersion. They expressed a Raman process, in which an atom goes from an initial level n to an intermediary level k, and then finally to a last level m. Each transition is described by a matrix element, and the sum over the intermediate states is equivalent to a matrix product. Heisenberg was inspired by this kind of interpretation, and he proposed other ones: for example, any matrix product could be described as a series of physical processes, even if they were only imaginary processes. (This mental representation of mathematical operations reappeared later in the virtual processes of quantum field theory.)

The abstract character of matrix mechanics was nevertheless startling, and it had not been anticipated by its discoverers. Three centuries of classical physics had made commonplace the identification of a physical concept with a number. Old questions about the reduction of physics to a purely quantitative construction were forgotten even though they were considered as nonobvious by Galileo and Descartes and as very questionable by many of their contemporaries.

Paul Dirac investigated the meaning of matrix mechanics by returning to the notion of a physical quantity. He took for granted the existence of real, physical "objects," whether they were electrons, protons, photons, or particles to be observed later. He also took for granted that they could be described by purely mathematical thought "objects." Drawing a lesson from Heisenberg's matrices, he recognized what was from then on the "hard core" of quantum mechanics: the description of a particle by mathematical objects that are not simple numbers. Dirac called them q-numbers, of which Heisenberg's matrices were one example and, later, the operators in wave mechanics another. They are not ordinary numbers (c-numbers in Dirac's language) because they do not commute. Their link with experiments lies in some numerical values, c-numbers, which are already contained in a q-number and which are accessible to a measurement. The matrix representing an atom's energy has for instance a set of eigenvalues: the c-numbers one can get from an experiment.

To conclude the discussion of the Born-Heisenberg-Jordan theory, one may notice that it was still lacking an essential notion: the "state" of a system. The new quantum mechanics could describe many features of a physical object, such as the energy levels and their quantum numbers. The state of that system, on the other hand, namely its exact situation at a given time, had not yet been found. It would come from wave mechanics. One may guess that, sooner or later, questions would arise about the trajectory of an electron in a detector and certainly whether there were vectors on which Heisenberg matrices could act, as any genuine matrix does. State vectors would have then come out with their physical meaning, and also their relation with wave functions, or at least the functions already known from the investigations of the mathematicians David Hilbert and Erhard Schmidt. The course of events was however going too rapidly and it took another direction.

4

Wave Mechanics

DE BROGLIE'S WAVE

28. I have already mentioned the interest that specialists had in the straight-line trajectories of X-rays. The existence of photons had been confirmed by Arthur Compton in 1921, and Louis de Broglie, who was familiar with the X-ray community, was soon to ask a rather simple question: if light, whose wave characters are known, can exhibit the properties of a particle, could it be that a wave is also associated with a genuine particle?

De Broglie investigated the possibility of a wave being associated with a free electron. His main result was a relation between the wave number k and the momentum of the electron p, namely,

$$p = \hbar k, \tag{4.1}$$

that is, a relation $\lambda = h/p$ between wavelength and momentum.

He obtained this result by considering a relativistic change of reference frame, or two different observers. The first one sees the electron at rest, and the wave (a "wave packet") is localized in the vicinity of the particle. The wave frequency is related to the electron energy by Einstein's rule for the photon $\nu = E/h$. The rest energy of the electron is $E_1 = mc^2$, so that $\nu_1 = mc^2/h$. A second observer sees the electron moving at a velocity v, with an energy $E_2 = mc^2/\sqrt{1 - v^2/c^2}$ and a wave frequency E_2/h. The four quantities $(\omega = 2\pi\nu, k)$ are the components of a four-vector so that, in the second reference frame, the wave number is $k_2 = v\omega_1/(c^2\sqrt{1 - v^2/c^2})$. This is equivalent to equation (4.1). De Broglie was convinced of the possible existence of the wave from the fact that the wave packet velocity (the group velocity $\partial\omega_2/\partial k_2$) is equal to the electron velocity v, so that the wave goes along the moving electron.

De Broglie also noticed a suggestive relation with Bohr's quantization rule for the hydrogen atom, $\int p\, dl = nh$. It looked as if a stationary wave could be standing on the electron trajectory, the length of the trajectory being an integral multiple of the wavelength.

What about Experiments?

29. If there is a wave accompanying an electron, it might show up in interference or diffraction effects. Some experimental data were suggestive of such effects, but they were not really convincing. It was only from an experiment by Clinton Davisson and Lester Germer in 1927 that diffraction of an electron could be seen as very similar to the diffraction of X-rays.

A crystal is a regular lattice of atoms. When a plane electromagnetic wave arrives at a unique atom, it is scattered, and a spherical wavelet centered on the atom is produced. In a crystal, there are many such wavelets and they interfere. In the case of X-rays with a wavelength of the order of the atoms spacing, interferences are destructive in most directions and constructive in a few Bragg directions, along which all the wavelets have a coherent phrase. Outgoing X-rays are concentrated along these directions, which only depend on the X-rays wavelength, crystal orientation, and the atom lattice.

Davisson and Germer's experiment exhibited similar diffraction effects for electrons crossing a crystal, in the same directions as would be obtained from a wave with a wavelength given by equation (4.1). As a matter of fact, the experiment was done at a time when it could only be a confirmation, because much progress had been made meanwhile on "matter waves," as we shall see.

SCHRÖDINGER'S EQUATION

Schrödinger's Eigenvalue Equation

30. De Broglie's idea could be used only if one knew the dynamics of a wave—how it evolves in time in general. Erwin Schrödinger had the necessary background for solving the problem. He had a good knowledge of mathematical physics, particularly the eigenvalue equations from which one computes resonance frequencies in electromagnetism and acoustics. This was exactly the kind of problem he encountered first.

De Broglie had only considered a wave associated with a free particle, and his remark on the hydrogen atom, although suggestive, did not provide a knowledge of the wave associated with a bound state. The latter was the problem Schrödinger wanted to solve. Starting from de Broglie's relativistic approach, he first constructed a relativistic theory for the hydrogen atom, but it did not work: the spectrum fine structure did not fit the data (because there was no relativistic theory of spin). He then tried a nonrelativistic approach and obtained an eigenvalue equation for the energy levels. Now the results were correct.

The title of Schrödinger's first paper, published in January 1926, makes things clear: "Quantization As an Eigenvalue Problem." He considered the wave $\psi(q)$ associated with a classical system whose classical energy (the Hamilton function) is $H(q_k, p_k)$. Using the correspondence principle together with some considerations in higher classical mechanics (the Hamilton-Jacobi equation; one may remember that Hamilton discovered his famous equations of motion from an analogy between mechanics and wave optics. Schrödinger essentially followed the same track in the opposite direction), he obtained as a reasonable guess an equation for ψ in the case of a stationary state with energy E. This is the *Schrödinger equation*:

$$H\left(q_k, \frac{\hbar}{i}\frac{\partial}{\partial q_k}\right)\psi(q) = E\psi(q). \tag{4.2}$$

In the case of a free electron, the coordinates $\{q\}$ reduce to the position x and $H = p^2/2m$. One has therefore

$$H\left(x_k, \frac{\hbar}{i}\frac{\partial}{\partial x_k}\right)\psi = -\frac{\hbar^2}{2m}\Delta\psi,$$

with Δ being the Laplacian operator $\partial^2/\partial x^2 + \partial^2/\partial y^2 + \partial^2/\partial z^2$. If ψ is a periodic function (a sine, cosine, or complex exponential) with a wave number given by equation (4.1), the Schrödinger equation (4.3) is satisfied with $E = p^2/2m$. One has therefore obtained a convenient generalization of de Broglie's initial hypothesis.

The Hydrogen Atom

31. In the case of the hydrogen atom, equation (4.2) becomes

$$-\frac{\hbar^2}{2m}\Delta\psi + V(r)\psi = E\psi, \tag{4.3}$$

with a Coulomb potential $V(r) = -Ze^2/r$. Schrödinger had no difficulty in solving this equation. Rotation invariance suggested introducing spherical coordinates (r, θ, ϕ). Schrödinger knew that the Laplacian operator Δ acts very simply on spherical harmonics $Y_{lm}(\theta, \phi)$, which had been known for a long time and depend on two integers (l, m). Looking therefore for a solution $\psi = f(r)Y_{lm}(\theta, \phi)$, Schrödinger obtained a differential equation for the radial function $f(r)$:

$$-\frac{\hbar^2}{2m}\left(\frac{d^2f}{dr^2}-\frac{l(l+1)}{r^2}f\right)-\frac{Ze^2}{r}f=Ef.$$

He knew the solutions of this equation, which is a special case of the so-called confluent hypergeometric equation introduced previously by Karl Friedrich Gauss. The asymptotic properties of the solutions for large values of r can be used to obtain the relevant values of the energy E. The function $f(r)$ increases exponentially for large values of r, except for some special values of E (the eigenvalues) for which, on the contrary, it decreases exponentially. In the later case, the function $\psi(x)$ represents a wave localized in a neighborhood of the nucleus, that is, in a bound state. The energy levels obtained in that way agreed with the data.

Schrödinger confirmed his results with a few more investigations. The corrections coming from the effective mass (see Section 12) were obtained when the electron and the nucleus are both associated with a wave function. The Zeeman and Stark effects were explained quantitatively by a perturbation calculation. Finally, Schrödinger wrote out an eigenvalue equation for an atom consisting of several electrons, from which he derived a few properties.

Dynamics

32. In his last article published in June 1926, Schrödinger wrote out an equation for the time evolution of an arbitrary wave function $\psi(q,t)$. Using again some analogies with classical mechanics and the correspondence principle, he obtained the basic equation

$$i\hbar\frac{\partial\psi}{\partial t}=H\left(q_k,\frac{\hbar}{i}\frac{\partial}{\partial q_k}\right)\psi. \tag{4.4}$$

Insofar as the wave function is supposed to determine all the observable properties of a system, this equation is the new form it must take for the fundamental law of dynamics. It may be noticed that the explicit use of complex numbers becomes compulsory in that case, whereas the imaginary quantity i could be disposed of in practice in the stationary equation (4.3). The time-dependent wave function for a stationary state, for instance, now contains a complex exponential factor $\exp(-iEt/\hbar)$.

WAVE FUNCTIONS AND PROBABILITIES

33. It was clear after Schrödinger's discoveries that the wave function provided a major key of quantum physics, but its meaning remained obscure. For a short time, Schrödinger could even believe that the "aberrations" of matrix mechanics would disappear and one could return to a simple, intuitive physics.

He interpreted the dynamical equation (4.4) by assuming the existence of an "electronic fluid" rather than pointlike electrons. The mass density of the fluid was supposed to be given by $\rho(x) = m\psi^*(x)\psi(x)$, and there was a mass flux $J(x)$ with components

$$J_k = \frac{m\hbar}{2i}\left(\psi^* \frac{\partial\psi}{\partial x_k} - \frac{\partial\psi^*}{\partial x_k}\psi\right). \tag{4.5}$$

These two quantities are related by an equation for the conservation of mass, as in fluid mechanics

$$\frac{\partial\rho}{\partial t} + \operatorname{div} J = 0, \tag{4.6}$$

which is a consequence of the dynamical equation (4.4).

Max Born had a quite different interpretation of the wave function, which he published in July 1926. He had immediately realized that the main novelty in de Broglie and Schrödinger's approach was a description, through the wave function, of the state of a system. This was the missing ingredient in matrix mechanics, and Born therefore considered wave mechanics as an essential complement to matrix mechanics and not a reason for rejecting it. (Dirac and Schrödinger himself would soon show the equivalence between the two forms of quantum mechanics). Born had been in search of a description of states, and he expected to get it by understanding scattering phenomena. One may notice that scattering had been at the origin of the whole story since it gave rise to the Rutherford model for atoms, and Bohr maintained that most questions of principle arising in quantum mechanics have a counterpart in scattering theory. Born was thus ready to test the concept of wave functions in this framework.

Born considered a plane de Broglie wave representing a free particle with a definite momentum. The particle hits a heavier scatterer or, equivalent, the wave is scattered by a fixed obstacle. Wave scattering had already been investigated thoroughly in the case of electromagnetic and acoustic waves, and Born, whose knowledge of physics was remarkably wide, knew the corresponding mathematical techniques. He showed that the wave, after scattering, is a linear superposition of many plane waves, each of them representing the scattered particle with a definite momentum. At this point, with no further discussion, Born bluntly asserted that only one interpretation was possible. The wave function $\psi(x)$ gives the probability for the particle to be detected at the position x. In a footnote, he added that this probability must be proportional to $|\psi(x)|^2$ (Born, 1926).

The meaning of the wave function is therefore as follows. The quantity $\rho = \psi^*\psi$ has nothing to do with a fluid, and it gives the probability density for the particle to be at the various points of space. More precisely, the probability p for the particle to be in some space region V at time t is given by

$$p = \int_V |\psi(x,t)|^2 \, d^3x, \tag{4.7}$$

assuming the wave function to be normalized by the condition

$$\int |\psi(x,t)|^2 \, d^3x = 1, \tag{4.8}$$

where the integral extends over all space (this normalization condition ensuring that the total probability is 1).

Born's introduction of probabilities in quantum mechanics reveals an essential feature of quantum physics that had been practically ignored until then. He was aware of the vast consequences of this idea, as shown in his conclusion: "Here the whole problem of determinism comes up.... I myself am inclined to give up determinism in the world of atoms. But this is a philosophical question for which physical arguments alone are not decisive" (Born 1926, p. 54). With these words, the era of interpretation had begun.

PART TWO

A Short History of Interpretation

When following the genesis of quantum mechanics, we saw how it inevitably became a new kind of science. Its outcome was a formal science, estranged from common sense, whose main concepts have lost a direct contact with intuition and have become mathematical symbols. Never before was there a science whose understanding required another special science: interpretation. No previous theory, moreover, had such strong internal tensions—between formalism and reality, randomness and causality—and paradox—linear equations whose linearity disappeared at a large scale. The uncertainty relations rejected as impossible a vision of the atomic world, and the most acute minds realized that the foundations of knowledge had drastically changed.

I will keep to my historical approach when introducing interpretation, even though the main theme of this book is to describe it as it stands now. There are many reasons for adopting a narrative approach rather than a didactic one: because new ideas are often the outcome of older premises; because the main problems were identified by some of the best people in the entire history of physics and it is better to quote them than pretending to start wholly afresh; and finally because the deepest features of the new interpretation are much better understood in the light of old primary questions.

There were always two main trends in interpretation, with Bohr and Einstein their two spokesmen. It seems that one can fairly reduce their opposition to a divergence on a unique question: is quantum mechanics complete? If not, the process of finding the right principles of the theory is not yet achieved and it must continue. On the contrary, if the theory is complete, there should appear a sound logical consistency in the interpretation. Most other controversial issues in quantum physics hinge on this one; this is true, for instance, of the status of reality: one cannot begin to discuss it before deciding on the issue of completeness.

Bohr, Heisenberg, and Pauli considered quantum mechanics to be complete. They realized that this belief implied a breach with the traditional, classical philosophy of knowledge. The opposite school pleading for incompleteness was led by Einstein and Schrödinger. Through a magnificent dialectical exchange, with both sides dedicated to the search for truth, they left the next generations with a few well-defined questions. It is telling, also, that no further hint of incompleteness has yet been found in spite of a much increased knowledge.

Our account of this history will be centered on four specific problems, which have evolved significantly since the origins:

1. The first one is about language. Can there be a universal language for interpretation? Can one speak with the same words, under the same constraints of logic, of an atom and of the commonplace world we can see and touch? This question was perhaps answered too soon by the idea of complementarity.
2. The second problem deals with quantum-classical correspondence. Is classical physics a direct consequence of quantum mechanics? The main difficulty of this question is not so much to find a mathematical correspondence between the Schrödinger equation and the classical laws of dynamics but to reconcile quantum probabilism and classical determinism.
3. The third question is the famous problem of Schrödinger's cat. The mathematics of quantum mechanics are linear, implying a universal occurrence of interference effects. But no such interference is ever seen at a macroscopic level, and this nonoccurrence is even essential for the existence of definite physical facts. How can one get out of this fundamental contradiction?
4. Finally, how are we to understand quantum mechanics? Is there a well-defined epistemological approach through which one can see quantum mechanics (and therefore most of physics) as a *clear*, fully consistent, and complete body of knowledge?

The history of interpretation will be divided into three parts, beginning in Chapter 5 with the Copenhagen version. Chapter 6 deals with the tumultuous period afterward and the contributions by Einstein, Schrödinger, von Neumann, de Broglie, Bohm, and Bell to name but a few. Chapter 7 will then sketch which answers have been found recently to the four key problems. They will be developed in more detail in Part Three.

5

The Copenhagen Interpretation

Quantum mechanics had hardly been discovered when worries about interpretation began to appear. The simplest definition of interpretation would say that it consists in understanding the theory. Bohr, Heisenberg, and Pauli recognized its main difficulties and proposed a first essential answer. They often met in Copenhagen, where Bohr had his laboratory, whence the name *Copenhagen interpretation.*

THE DIFFICULTIES OF INTERPRETATION

Quantum Physics and Mathematics

34. Abstractness was certainly the most startling feature of the new theory. What are these complex wave functions? What does it mean to replace genuine dynamical variables by infinite matrices or operators? When Schrödinger and Dirac established the equivalence of matrix and wave mechanics, these difficulties were there to remain, and people spoke for some time of a "symbolic" physics, by which they meant the symbolism of mathematics. The new physics was *formal*, meaning a science whose basic notions and fundamental laws can only be fully expressed in a mathematical language, using symbols in place of physical concepts.

Bohr, Heisenberg, and Pauli saw this precisely as a crisis of concepts, a word I will use with the same meaning as they did. A concept is supposed to be very rich, with many meanings and many associated ideas; it has been nourished by many perceptions and multitudinous examples, as illustrated by the concepts of object, substance, space, time, motion, cause, etc. The abundant content of a concept was opposed to the ascetic aridity and definiteness of a mathematical symbol (or "notion"). For the Copenhagen thinkers, one of the main problems of interpretation was the impoverishment of a physical concept when it is reduced to a mathematical symbol, the more so when the cognitive distance between the concept and the symbol is great, as between a physical quantity and an operator.

35. There had been already some years beforehand another formal science, namely relativity and its later extension to the relativistic theory of gravitation. There also an interpretation had been necessary, as an addendum to the theory itself. Its aim was to relate the mathematical symbols to observable data. The task had been rather easy because it was enough to introduce various observers in different reference frames and to translate the mathematics of the theory by their observations.

The trick did not work for quantum mechanics. There was no way of translating its mathematical symbols into some commonsense intuition of an observer. This conflict with common sense raised a huge difficulty and was in opposition with everything mathematics had previously been able to do in physics. In classical physics, mathematical symbols can make physical concepts clearer and more precise. When a position, a velocity, or a law of dynamics is expressed by some coordinates, derivatives, and differential equations, the mathematical symbols add to the clarity of the physical concept. There is a full agreement between the theoretical tool and the reality it stands for. Little of this simplicity remains, however, in quantum mechanics. The precision and consistency of mathematics are still there, but their relation with intuition is almost completely lost.

Complex numbers are a good example of these difficulties. These numbers had been used occasionally in classical physics but only for making a calculation simpler, for instance, when an oscillating signal is considered as the real part of a complex exponential. Complex numbers are superfluous, and every calculation can be done without them. It is no accident if the same word "real" is used for real numbers and physical reality: the concept of reality calls for symbols that are real numbers. How can it be that quantum mechanics requires complex ones?

A simple example may show that one cannot generally use real wave functions because of their probabilistic meaning. Consider a wave function for a particle having a definite momentum p in some reference frame. It must be a periodic function according to de Broglie, and if real, it can be written as $A\cos[(px - Et)/\hbar + \alpha]$. At time zero, the probability for observing the particle at a point x vanishes at the nodes of the cosine, which are separated by a distance $\lambda/2 = \pi\hbar/p$. In another reference system moving at a velocity V with respect to the first one, the momentum is $p' = p - mV$, and the points where the probability of observation vanishes are separated by a different distance $\lambda'/2 = \pi\hbar/p'$. This is obviously impossible, since distances are invariant in nonrelativistic physics. The difficulty disappears when the wave function is a complex exponential.

Absolute Randomness

36. Born had anticipated serious difficulties when he proposed his probabilistic interpretation of the wave function: if the wave function is a fundamental notion of the theory, it implies some sort of universal randomness in the world of atoms.

Probabilities had appeared previously in the "old quantum theory" with the quantum jumps of an electron going from an atomic trajectory to another one. Einstein had also introduced probabilities in the emission of black-body radiation. In both cases however, it had been thought that probabilities were needed for the same reasons as in classical physics, simply because all of the laws were not yet known. Now, however, with a full-fledged theory of quantum mechanics, absolute randomness was plainly there.

Quantum randomness had little in common with its classical counterpart, and Born was fully aware of that difference when he raised the problem of determinism in his memorable paper. Determinism is an intrinsic part of classical physics (even if some care must exercized in the case of chaos). The differential equations of classical dynamics predict exact values for position and velocity at any time when they are given at an initial time. Random events do not, therefore, exist in classical physics. Randomness is an appearance originating from the ignorance of initial conditions or from the inability to measure them or to compute explicitly the motion. These three kinds of limitation occur together when one throws dice, and this is why the result is said to be random, although it is not in principle.

The situation is quite different in quantum mechanics. One might know perfectly well a plane electron wave arriving on an atom. One might also, in principle, compute the wave function exactly at a later time by solving the Schrödinger equation. Everything entering into the theory is thereby exactly known yet nevertheless, one cannot predict which detector will react to the scattered electron. One is facing an intrinsic form of probabilism having no ignored or hidden causes: absolute randomness. One knows Einstein's famous rebuttal that "God does not play dice."

Probabilism versus Determinism

37. The difficulty with absolute quantum randomness is made still more acute when one notices that there would be no experimental physics if determinism were forsaken. This is a remarkably strange circumstance for an experiment dealing with quantum processes because verifying a

purely probabilistic theory requires macroscopic determinism, even though the instruments are made of atoms! But a good experimental device must be deterministic: it must work according to the directions for its use, as one expects from a car or a washing machine. This is ordinary determinism, in the usual direction of time. When I push a particular button, the device starts exactly as I expected. If not, I cannot rely on it and particularly not use it as a reliable part of a scientific instrument.

Quite remarkably, determinism enters also in another guise in the verification of quantum predictions. A prediction from quantum mechanics is most often a probability, and in order to check it, one must perform a long series of measurements. Comparing how many times a specific event actually occurs with its theoretical probability, one can check the prediction. It may be noticed however that the results of the various measurements must have been recorded previously—either in a notebook, a computer memory, or a photograph. But using a record as a faithful witness of a past datum is another instance of determinism, although in the opposite temporal direction, where the present existence of a record stands for the validity of a past event.

This example also suggests how probabilism and determinism could be reconciled. Records are not perfect; they can be damaged, and even in the best conditions of conservation, they are exposed to quantum fluctuations: molecules in the dots of ink on a document can move for instance, although the probability for a strong fluctuation is extremely small. Determinism could therefore enter in the framework of probabilism, an event being determined when the probability for its nonoccurrence is very small.

This is an example of an idea that appeared long ago, probably first in Heisenberg's papers, and matured a long time before gaining recognition. It evolved slowly from a hint, lost among other conflicting ones, to a full proof. Its course of development shows the futility of identifying a single origin of an answer to the basic questions of quantum mechanics, because many of them began as suggestions without much conviction backing them up. Also, many were proposed several times independently. I therefore will not try to follow in detail the history of the ideas that proved fruitful. This task must be left to historians.

The Correspondence Principle

38. The two difficulties I mentioned concerning mathematical symbolism and randomness point toward another problem of interpretation, which is how to take into account two opposite aspects of physical reality. One of them is in the laboratory and all around us, a macro-

scopic, causal, and intuitive world. The other dwells in atoms and particles, symbolic in its description and subjected to pure randomness. The difficulty is that these two are only one real world.

We saw how Bohr insisted on the existence of a continuity between the two domains or the two visions of the world. This was expressed by the correspondence principle, although Bohr was never able to state it explicitly like a rule one can apply in a straightforward way. To say grossly that classical and quantum physics should agree, at least when Planck's constant can be considered as small, seems rather obvious, or it should be. But to make this statement explicit and precise is much more difficult. One may compare this question with a similar one in relativity theory: the limit $c \to \infty$ can be worked out easily, but with the limit $h \to 0$, one is at a loss to say how an operator tends toward a number and how one eliminates wave functions at a macroscopic level. Correspondence is certainly much more than a simple limit and is one of the deepest problems in quantum physics.

UNCERTAINTY RELATIONS

39. The first tangible result of the new interest for interpretation was the discovery by Heisenberg in 1926 of the uncertainty (or indeterminacy) relations. Because the wave function can be used for obtaining probabilities, the average value and the statistical uncertainty of a physical quantity can be defined according to probability calculus. Heisenberg considered the uncertainties Δx and Δp_x for the x-components of the position and momentum of a particle. In the case of a Gaussian wave function, he found the product $\Delta x.\Delta p_x$ to have the value $\hbar/2$. Wondering about the generality and the meaning of the result, he devised various imaginary (*gedanken*) examples, among which is the famous "Heisenberg microscope."

Heisenberg's microscope. An electron is located approximately in the focal plane of a microscope, which can work with a light of arbitrary wavelength (see figure 5.1). One wants to measure the x-component of the electron position (the x-direction being in the focal plane). A monochromatic light with a wavelength λ is used for this purpose. When looking through the microscope, one can find the value of x, although there is an unavoidable uncertainty Δx because of the finite resolving power of the instrument. Because of the photon structure of light, there is also a limitation on the precision with which the electron momentum can be known. Although several photons are needed for obtaining an image of the electron, one may consider the effect of a

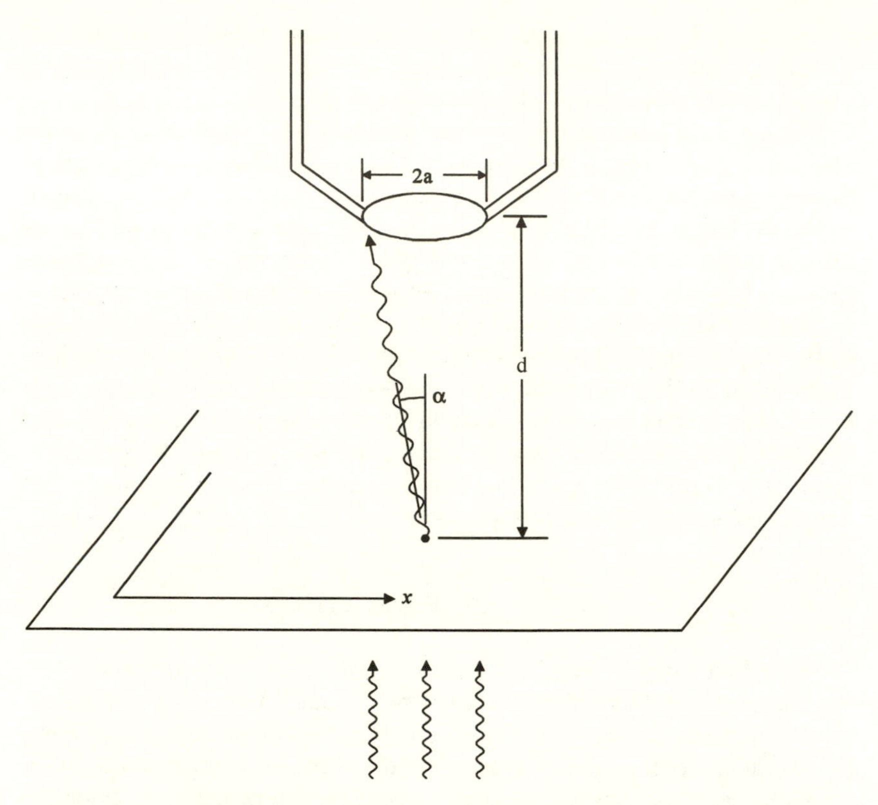

Figure 5.1. Heisenberg's microscope. Collisions of photons with an electron result in an uncertainty Δp_x in a component of the electron momentum. Diffraction limits the resolving power of the microscope, implying an uncertainty Δx in the x-coordinate of the electron position. The Einstein relation between a photon momentum and the radiation wavelength implies an uncertainty relation between Δx and Δp_x.

single photon. Initially it has a definite momentum, and when scattered, it contributes a recoil momentum to the electron's momentum. One does not know the recoil momentum of the photon exactly but only that the photon went through the front lens of the microscope. Because of the finite size of the lens, there is an uncertainty in the momentum of the scattered photon, implying an uncertainty Δp_x in the x-component of the electron momentum. Heisenberg could then show that the product $\Delta x . \Delta p_x$ must be of the order of h.

The calculation is easy. Let d be the distance from the focal plane to the front lens, the radius of which is a. One assumes $a \ll d$ for simplicity. The lens is seen from the focal plane under an angle $\alpha \sim a/d$. Let p be the initial photon momentum along the microscope axis. The uncertainty Δp_x of the x-component of the scattered photon momentum (which is the same as the uncertainty on the electron momentum) is $p\alpha$, with $p = h/\lambda$. If $\Delta\theta$ is the resolving power, the uncertainty on the position is $\Delta x \sim d.\Delta\theta$. Finally, one relies on diffraction theory for an estimate of $\Delta\theta \sim \lambda/a$, that is, the resolving power. This gives for the product $\Delta x.\Delta p_x$ a value of the order of h.

The uncertainty Δp_x results from an indeterminacy in the photon momentum, that is, from considering light as composed of particles, whereas Δx is due to a finite resolving power under diffraction, that is, from light considered as a wave. This shows a close relation between the uncertainty relations and wave–particle duality.

Bohr's ideal experiment. Bohr also analyzed the uncertainty relations by means of various ideal experiments, particularly the following one: An electron goes through a slit in a screen, which has a width Δx. If the electron is observed behind the screen, one knows the value of its x-coordinate at the time when it crossed the screen with an uncertainty Δx. If the electron arrives in the direction normal to the screen, its initial momentum component p_x is zero. The momentum could have changed by a quantity Δp_x when the electron crossed the slit because of an interaction with the screen. One can reach this quantity by measuring the recoil momentum $\Delta p'_x = -\Delta p_x$ of the screen. For $\Delta p'_x$ to be measurable, however, the screen must be free to move in the x-direction. It must have a position uncertainty $\Delta x'$ not larger than the slit width Δx, if Δx is to be considered as the correct error in the electron position. If one applies quantum mechanics to the screen, the product of the corresponding uncertainties $\Delta x'.\Delta p'_x$ is of the order of h, and this implies a similar relation for the electron.

Bohr's argument is important because it shows the consistency of the uncertainty relations. It also implies that a macroscopic object, however big, must be sensitive to quantum limitations.

It was proved later that the lowest bound for the product $\Delta x.\Delta p_x$ is $\hbar/2$, as obtained with a Gaussian wave function. Other uncertainty relations were described for any pair of noncommuting observables by means of their commutator.

The most important consequence of the uncertainty relations for interpretation is their incompatibility with an intuitive representation of a particle as being a point in space. The idea of a space trajectory is also excluded because it would mean simultaneously precise values for position and velocity. The "concept" of particle becomes obviously much poorer.

A PHILOSOPHICAL REVISION

40.* (*Note*: Some sections of this book can touch on somewhat technical questions and they may be most often omitted in a first reading. They are indicated by an asterisk. The technicalities (?) in the present section are those of philosophy.) Bohr, Heisenberg, and Pauli had realized that the interpretation of quantum mechanics would need much more than a straightforward translation of the formalism into some observation procedures, as had been done for relativity. They recognized that much of the traditional philosophy of knowledge had to be challenged, down to its very foundations.

This aspect of interpretation has been investigated most thoroughly by Pauli, first in private letters and later in articles in which he appears as a physicist and as a philosopher. As a physicist, he recalled the genesis and evolution of physical concepts from Kepler to Newton and compared them with the demands of quantum mechanics. He also questioned the classical philosophy of knowledge from a modern standpoint, and this part of his work is extremely interesting for anyone interested in interpretation.

Kant's *Critique of Pure Reason* being the epitome of the classical philosophy of knowledge, Pauli proceeded to a careful examination of its preconceptions. One of Kant's basic pronouncements is that space and time are essential modes of our thinking: we absolutely cannot think of the mutual relations between different objects except by seeing them definitely in space and continuously in time. If this were true, it would preclude a probabilistic location of a particle in space and contradict the uncertainty relations and their incompatibility with definite trajectories.

Another essential feature of Kant's theory is the existence of categories of understanding. A category is a framework of reason from which our understanding cannot escape as a matter of principle. Kant had carefully derived twelve such categories from a systematic analysis and thereby provided later generations with a clear definition of what might be called the classical theory of knowledge. Some aspects of the construction were later questioned by philosophers, but never the essential aspects of the categorical judgements about space and time nor the categories of the understanding. Pauli was the first to do so. He called attention to the fact that one of the categories, causality, is obviously at variance with atomic physics. How can then one accept it as a necessary condition for reason if it is not necessary for Nature?

Another problematical category is reality. Quantum mechanics agrees with the existence of a reality that is independent of mind, but Kant asked for more in his transcendental analytic, which is supposed to be a complete dissection of knowledge into distinct elements. It means that something real cannot be described by two concepts excluding each other. Something real cannot be a bird and a tree, a stone and a sound, a wave and a particle.[1]

Without delving too much into philosophical matters, one can easily summarize the outcome of Pauli's investigations: The aim of an interpretation of quantum mechanics cannot be limited to an agreement between the abstract theoretical formalism and common sense, which applies in the macroscopic world of experience. It must involve, in one way or another, a radically new manner of thinking and understanding.

COMPLEMENTARITY

41. A most interesting feature of the first attempts at an interpretation by Bohr, Heisenberg, and Pauli, was their insistence on finding a language that would be specific to interpretation. The idea of a language implies some universality, that is, a possibility of expressing the whole of physics. A language, moreover, cannot be dissociated from the underlying logic giving it unity. Bohr, however, changed the emphasis in a famous lecture he gave in a congress held in Como in 1927. Rather than a universal language of physics, he suggested that complementarity was the essential logical character of a quantum language. Complementarity was for him a universal and fundamental principle, although it always remained somewhat unclear, like the correspondence principle.

The complementarity principle states that mutually exclusive modes of language can be applied to the description of an atomic object, although not simultaneously. A few examples are, in fact, clearer than a general rule. We may speak of an electron by using the language of waves when it crosses two slits in an interfering device, and we may

[1] Bernard d'Espagnat has investigated the problems arising from the notion of reality, particularly in terms of a "veiled reality." This is a physical reality, existing in itself, accessible to physical investigations but which is not in accordance with Kant's category of reality, that is, with the kind of conditioning our mind would like to give it. See d'Espagnat (1979, 1994).

speak of the same electron as a particle when it is detected, but we cannot use the two modes of speaking at the same time. When using the language of particles, we may choose, moreover, to speak of the position at a given time or of the momentum, but, again, not both together.

The choice of the right language is determined by the real experimental device one is considering. A screen with two slits implies the language of waves, and a detector calls for the language of particles. The essential role of the means of observation led Bohr to bring forward a notion of classical philosophy, that of "phenomenon," meaning something that can be directly perceived (according to Kant). In Bohr's formulation of the language of interpretation, the only data at our disposal are given by an experimental device. They are phenomena, if the role of perception is extended to cover an observation with the help of an instrument. Finally, what is observed determines the language to be used: phenomena dictate their own description.

The idea of complementarity is deep and subtle. It considerably limits our representation of reality, which ceases to be permanent and universal and is split into a multitude of successive situations, each one expressed by an extremely reductive language (something can be said for instance to have a position, or to have a momentum). Complementarity also introduces a deep ambiguity into the concept of object, which can look like a wave or a particle in different circumstances. No wonder then if the language expressing atomic events becomes confusing and apparently irrational, as it would appear if transported into our surrounding world with a sentence such as: "This man speaks pleasantly on the phone, but he is always a cat when one can see him." Most physics textbooks try to avoid this tricky question whereas philosophical texts delight in it. Bohr himself saw complementarity as a philosophical breakthrough, and he proposed extending the idea to physiology and psychology, although it seems difficult to see more in these attempts than analogies and metaphors.

42. It was probably unfortunate that the search for a universal language of interpretation was soon replaced by Bohr's insistence on complementarity. As a matter of fact, the first elements for a truly universal language of interpretation were discovered by John von Neumann in 1929 and 1930, as will be seen in the next chapter. His *elementary predicates* were the intermediate link needed between the usual language used for the description of experiments and the abstract theoretical formalism. The idea unfortunately ran up against two kinds of problems: macroscopic interferences, which were to become famous with the example of Schrödinger's cat, and difficult problems in logic. The two obstacles persisted for years, so von Neumann's proposal did

not look fruitful at the time, and the Copenhagen trio did not pay much attention to it.

In any case, interpretation could not wait long because it was needed in the everyday work of experimentation. Opportunities were opening in chemistry and physics, and they could not be delayed or deemed doubtful because of some philosophical scruples. Bohr therefore proposed the necessary rules. He was particularly aware of the logical difficulties one could encounter when using interpretation. He therefore chose to insist on the notion of truth as the most essential aspect of logic in science. In a natural science, truth relies primarily on facts, which are observed phenomena. Truth can even be identified with facts and with the observed relations between facts. In view of this role of facts and also of the difficulty of speaking of quantum events because of complementarity, Bohr proposed to adopt the language of classical physics as the only reliable one.

43. Bohr's proposal agreed of course with the practices of experimental physics. Many theoreticians were also reluctant to engage in much philosophical questioning and were eager to embark on new applications. The idea was therefore easily adopted in spite of objections by Einstein, de Broglie, and Schrödinger and even though it had some serious inconvenience.

Two significant retreats had occurred from the initial search for a universal language: the withdrawal to the ambiguities of complementarity and the final entrenchment in a language of phenomena that was restricted to pure classical physics. Each step backward had been a loss of ambition for interpretation, and the result was two noncommunicating, untranslatable idioms. One was classical, commonplace, and well-suited to the world of experience. Its counterpart was a pure mathematical symbolism for the description of atoms that was hardly mitigated in its abstractness by a recourse to complementarity.

When a science relies on two languages—two different logical settings—there may be legitimate doubts about its consistency. The opposition between determinism and probabilism had never been finally settled and that was enough to raise serious questions. It was impossible, however, to put the problem of consistency to a test when the principles involved were the unclear ones of correspondence and complementarity. Nobody was probably more concerned with consistency than Heisenberg, who tried to complete the idea of correspondence by introducing a frontier between the domains of classical and quantum physics and their two languages. The frontier was more or less defined by the numbers of atoms under consideration. In the case of a Geiger counter detecting an electron, for instance, one can consider the counter itself with its electric signal as belonging to the classical domain. But one can also move the frontier so that a few ionized atoms represent the classical system, and the quantum electron is left to its mathematical description. Of course, many other intermediate steps are possible, and there is a large amount of freedom in the choice of a "frontier" with not much difference in practice. The idea was very sound, but it was only elaborated recently with the development of a quantitative test for the amount of determinism according to the frontier.

In spite of the importance of language in Bohr's work on interpretation, it is remarkable to see how little influence his final pronouncement had outside a small community of specialists with a philosophical inclination. Bohr maintained that only classical phenomena can be described by reliable concepts, whereas atoms and particles cannot. Most physicists, nevertheless, continued speaking freely of atoms and particles as if they could be seen and describing events occurring among them. Even worse, every textbook, every lecture on quantum mechanics, and most communications in scientific meetings, quietly ignored Bohr's warnings or paid only lip-service to them.

As Feynman (1965a) noticed, an interference experiment is always described as if the interfering object is always well defined, even though it is said to be a wave when it crosses two slits in a screen and a particle when it is detected. How can one understand a science that cannot even give a permanent name to its own object of study? "Nobody understands quantum mechanics," Feynman said elsewhere (Feynman, 1965c, p. 129) and perhaps this was the best summary of interpretation.

WAVE FUNCTION REDUCTION

44. When singling out strict classical physics for expressing experimental data, Bohr was creating new, deep problems. Because classical data are expressed by ordinary numbers, how can they be used for reaching a knowledge of a wave function? One must answer this question because one needs the wave function to use the theory and make predictions. The cogency of this question resulted in a great and perhaps excessive emphasis on measurements, which became the center of interpretation, so much so that interpretation for most physicists was reduced to a theory of measurements.

The theory of measurement is most often expressed by a set of rules that can be found in any textbook: (1) One can only measure the value of an observable, which is described by the theory as a Hermitian operator. The question whether every observable or only a few of them can be measured as a matter of principle is more controversial. (2) The result of an individual measurement of an observable A is a random event. (3) The values obtained belong to the spectrum of A. (4) In the simplest case, when this value a is a nondegenerate discrete eigenvalue of A with eigenvector $|a\rangle$, the probability for finding it is $|\langle a|\psi_i\rangle|^2$, where ψ_i is the wave function of the measured system at the beginning of the measurement.

These rules are followed by the rule of wave function reduction, which purports to give the wave function *after* the measurement. It is most easily expressed in the case of an ideal measurement, when the measuring device has the virtue of producing the same result a if the measurement is immediately repeated. According to rule (4), this means that the wave function ψ_f at the end of the measurement must be the eigenvector $|a\rangle$. This is the simplest statement of the rule of wave function reduction.

The validity of these rules is in no doubt. They have been checked by countless experiments in a wide variety of circumstances, and they often replace interpretation in the practice of physics.

45. The reduction rule, however, introduced a great difficulty in interpretation: Can one still speak of an interpretation of quantum mechanics when one has surreptitiously replaced it with another theory? The point is that quantum mechanics relies on an explicit dynamics, which is expressed by Schrödinger's equation. A measurement is only from this standpoint an interaction of a special type between two physical systems: the measured atomic object and the measuring device. One expects this interaction to obey a Schrödinger equation, even if a complicated one, but this expectation is contradicted by the reduction rule.

This contradiction can be seen plainly when the result of the measurement is a degenerate eigenvalue a. If P denotes the projection operator on the subspace of the eigenvectors $|a, j\rangle$ with eigenvalue a, the reduction rule gives for the wave function after measurement the expression

$$\psi_f = \frac{P\psi_i}{\|P\psi_i\|}, \tag{5.1}$$

where the denominator is necessary for normalization. Equation (5.1) is however a nonlinear relation between ψ_i and ψ_f, which cannot be an outcome of a linear Schrödinger equation. It can only mean that one has surreptitiously replaced Schrödinger's dynamics by something else, or so the critics said. What can be the meaning of an interpretation that is violating the theory it purports to interpret?

Bohr was fully aware of the difficulty. No doubt, he said, the reduction rule is only one of its kind in the whole realm of physics. It has no analogue anywhere because its role is to bridge the gap between two different physics, one of them classical, obvious, and the other symbolic, mathematical. The nonuniqueness of language was thereby reaching its most extreme consequences. Bohr did not waver at considering reduction as being a real physical effect. The violation of Schrödinger's linear dynamics was thereby ratified, accomplished through an external effect, which occurs only when there is a measurement.

But what is so special in a measurement among all possible interactions? Is it possible that reduction acts immediately at an arbitrarily large distance (as Einstein, Podolsky, and Rosen were to show soon afterwards)? What an incredible rule, as many people reflecting on it came to conclude, but also a perfectly correct rule because it works whatever may be the experiment to which it is applied. It is true because it works, and for no direct reason in the rest of quantum theory. It did not help much if reduction was considered as a transition from potentiality to actuality, an "objectification" from the set of all the theoretically possible results in a measurement to the unique one that is actually seen.

The last word was certainly not yet said.

6

Interpretation after Copenhagen

The name "Copenhagen interpretation" has not always meant the same thing to different authors. I will reserve it for the doctrine held with minor differences by Bohr, Heisenberg, and Pauli. We saw in the previous chapter why some serious problems led to questioning and even contesting this interpretation. The present chapter is a brief review of this dispute in so far as it contributed to a better appreciation of the underlying problems.

VON NEUMANN'S CONTRIBUTION

46. The *Mathematical Foundations of Quantum Mechanics*, by John von Neumann (1932) was a major contribution to interpretation. It is sometimes considered as belonging to the Copenhagen interpretation, but there are significant differences with the Bohr-Heisenberg-Pauli doctrine on issues that were later to become essential.

Contrary to Heisenberg and Pauli, who were intellectually close to Niels Bohr, von Neumann represented another line of ideas that originated with the mathematician David Hilbert. The latter had another conception of the nature of a physical theory. According to Hilbert and his disciples, a theory should rely on clear and explicit principles, that is, axioms. These founding assumptions must be clearly distinguished from their consequences, which must be inferred with perfect mathematical rigor. Much attention must be given to logic in questions dealing with foundations. Finally, logic itself is a formal science, which had been codified by Gottlob Frege and developed by Hilbert himself; it was nothing like the philosophical type of logic that was used by Bohr and even by Pauli.

This standpoint has important consequences for interpretation. The same theory cannot contain two categories of axioms, some of which belong to classical physics and others to quantum physics. Classical physics should be rigorously inferred from the quantum axioms in the macroscopic domain to which it belongs. Interpretation should be primarily a sub-theory within quantum theory rather than a commentary acquiring part of its inspiration from philosophical considerations.

None of these aims were accomplished rapidly, which may explain the lack of interest by the Copenhagen trio for these dissenting views. As for the physics community, it took from von Neumann's contributions principally the framework of Hilbert spaces and the difficulties of measurement theory.

The Hilbertian Framework

47. The mathematical theory of Hilbert spaces had been constructed at the beginning of the twentieth century for solving linear integral equations. Von Neumann recognized in them the most convenient framework for a precise formulation of quantum mechanics. Another framework had been previously proposed by Paul Dirac and can be found in his well-known book *Quantum Mechanics* (Dirac, 1930). The basic notions were wonderfully simple and invented directly for the sole purpose of quantum theory. They included the *delta function*, *bras*, and *kets*, terms that are still used today. A direct connection with the conventional body of mathematics was not made at the time and would come only much later. Dirac's mathematics therefore could not be considered as rigorous, and as a matter of fact, it sometimes involved slightly incorrect conjectures.

On the other hand, von Neumann's alternative proposal made possible a rigorous analysis of the foundations and applications of quantum mechanics that agreed on most points with Dirac's approach. A Hilbert space, for instance, is a vector space, and vectors can be added together or multiplied (by a complex number). This is a perfect setting for the linear Schrödinger equation or the superposition principle, which was considered by Dirac as the first principle of quantum theory.

Among all vector spaces, Hilbert spaces are distinguished by the existence of a scalar product $\langle \alpha | \beta \rangle$ between any pair of vectors α and β. There are various kinds of Hilbert spaces. Some have a finite dimension and their operators (acting linearly on the vectors) can be represented by matrices. This is particularly convenient for the description of spin states. A Hilbert space can also be infinite-dimensional, as well as the matrices representing the corresponding operators. One can then recover the infinite matrices, the idea of which had been introduced by Heisenberg, Born, and Jordan. Finally, the square integrable functions on a coordinate space also belong to a Hilbert space (denoted by L^2), which can be considered as composed of wave functions.

The mathematical theory of Hilbert spaces was still incomplete for the aims of physics. The article by von Neumann (1927) brought some decisive advances. The principal one was the famous spectral theorem, generalizing to operators a property of Hermitian matrices, which is the existence of an orthogonal basis of eigenvectors with real eigenvalues. Von Neumann defined self-adjoint operators generalizing Hermitian matrices (an operator A being self-adjoint if $\langle \alpha | A\beta \rangle = \langle A\alpha | \beta \rangle$ for any couple of vectors α and β). He also defined the spectrum of an operator (generalizing the set of eigenvalues), distinguished between discrete and continuous spectra, and showed how to deal rigorously with a continuous spectrum. Most physicists became annoyed with the mathematical complexity of quantum mechanics: they were offered a choice between von Neumann's Charybdus with his Lebesgue integrals and Dirac's Scylla with the delta function (which joined later the herd of conventional mathematics with the theory of distributions [Schwartz, 1950]).

Von Neumann (1940) also extended another of Dirac's ideas, namely the direct reliance on q-numbers. Von Neumann abstracted the algebra of operators in a Hilbert space by considering an a priori given algebra, or C*-algebra, which can represent observable quantities without being directly obtained from a definite Hilbert space. This approach is particularly useful in statistical mechanics, but in most cases it shows only that Dirac's q-numbers are essentially equivalent to the Hilbert space framework. Once again, quantum mechanics appears as a unique theory with many different faces.

Logic

48. I mentioned earlier that von Neumann had found the elements of a universal language of interpretation that had been first contemplated by Bohr, although he did not recognize it when it was shown to him. Von Neumann reduced the significant sentences entering into the discourse of physics to some elementary propositions and he investigated their relation with logic. This approach shows the strong influence of the logical school initiated by Hilbert.

Von Neumann's treatment of the basic propositions of physics, which he called *elementary predicates*, is particularly instructive. A predicate is a statement referring to an *observable* A, that is, a self-adjoint operator, and a set of real numbers Δ belonging to the spectrum of A. The predicate is expressed by the sentence "the value of A is in Δ." Von Neumann's brilliant idea was that one can associate a definite mathematical object with the predicate. This step was inspired by one of the versions he derived for the spectral theorem in terms of projection operators, and it turned out that a definite projection operator can be associated with a predicate. Von Neumann saw, furthermore, a relation between this statement and logic because the most basic property of a logical proposition is to be either true or false (viz. the principle of *excluded middle*). But one can symbolize "true" by the number 1 and "false" by 0, as shown by von Neumann's later decisive contribution to the logic of computers. The only possible eigenvalues of a projection operator are precisely 1 and 0, and this may explain the deep relation between projection operators and a logical language.

I will develop this analogy in greater detail in Part Three, but I must acknowledge that few arguments could back it up when von Neumann proposed it. There were at least two reasons. First, in spite of his attempts, von Neumann was unable to include the statements of classical physics in his new logical framework. There was therefore no sufficient evidence for the universality of his language. The second difficulty came from logic itself. If one considers all the possible predicates as being possible statements, there is no framework in which the basic operations of logic apply and agree with the axioms of logic. This is due once again to the noncommutativity of operators, namely of projection operators in the present case.

In a last desperate attempt, von Neumann made an extreme proposal in a paper written with George Birkhoff (1936). The classical and quantum modes of thought are there considered to be so wide apart that they do not obey the same rules of logic. Some of the standard rules originating with Aristotle and the Stoics or Boole and Frege would not be valid in the world of quanta. Such a proposal by two remarkable logicians is very strange. How could they expect that an agreement between the common sense of experimental physics and the abstractness of quantum theory would be obtained outside conventional logic, when both common sense and the logic of a mathematical theory are conventional? This is not the only occasion when we shall see a desperate attempt out of the maze of interpretation, and it shows the urgency and importance of these problems.

MEASUREMENT THEORY

The von Neumann Model

49. Another of von Neumann's important contributions to interpretation was a theory of quantum measurements. It appeared in the last pages of the *Mathematische Grundlagen*, which are reproduced in the source book by Wheeler and Zurek (1983). Von Neumann assumed that there was only one kind of physics: quantum physics. A measurement is therefore an interaction between two quantum systems Q and M. An

observable A is associated with the measured system Q and a second system M, the measuring device, is used to measure the value of A. Von Neumann considered a model where M has only one degree of freedom, representing the position X of a pointer on a ruler. The pointer is initially at position zero, which means in view of its quantum character that its initial wave function $\phi(x)$ is narrow and centered at $x=0$.

The interaction between the two systems Q and M is very simple. It is given by an interaction Hamiltonian $H_{\text{int}} = -g(t)A.P$, where A is the observable to be measured and P the pointer momentum canonically conjugate to X. The function $g(t)$ is large and positive during a short time interval Δt immediately after the time $t=0$, and it vanishes outside that interval. The interaction Hamiltonian is dominant during this short period so that the evolution operator for the whole system $Q+M$,

$$U(0,\Delta t) = \exp\left(-(i/\hbar)\int_0^{\Delta t} H(t)\,dt\right),$$

is practically equal to $\exp(-i\lambda A.P/\hbar)$, where $\lambda = \int_0^{\Delta t} g(t)\,dt$ can be made as large as desired.

If the initial state of the measured system Q is an eigenvector $|k\rangle$ of A with the eigenvalue a_k, the evolution operator acting on this state reduces to $\exp(-i\lambda a_k.P/\hbar)$. This expression can be recognized as the operator translating the pointer position X by a distance λa_k. The pointer wave function $\phi_k(x)$ after the interaction is therefore the same narrow function it was initially, now displaced to be centered at the point $x=\lambda a_k$. The new position of the pointer shows explicitly the input value of A, as one expects from a reliable measurement.

50. Things begin however to go wrong when the initial state of the measured system is not an eigenstate of A. We shall consider, for instance, a combination of two eigenstates

$$|\psi_o\rangle = c_1|1\rangle + c_2|2\rangle. \tag{6.1}$$

The initial state of the system $Q+M$ can be written as $|\psi_o\rangle \otimes |\phi\rangle$ and it becomes, immediately after the interaction,

$$|\psi\rangle = c_1|1\rangle \otimes |\phi_1\rangle + c_2|2\rangle \otimes |\phi_2\rangle. \tag{6.2}$$

There is now a striking difference between the physical properties of this state and ordinary experimental evidence. In a real measurement, one sees the pointer go to a definite position, which is either $\lambda.a_1$ or $\lambda.a_2$ according to a random occurrence. There is nothing similar in equation (6.2) where the two possible results stand on the same footing.

The linear superposition in the initial state (6.1) has been transferred to the measuring device so that possible interferences are now transferred to a macroscopic level. The difficulty cannot be removed if the measuring device M is observed (measured) by another apparatus M' and the contagion of interferences is caught by M'. This pattern, where a measuring device is measured by another measuring apparatus, which is measured, and so on, is called a *von Neumann chain.*

This model of a measurement by von Neumann had a great effect on interpretation. In contrast with the correspondence principle, it indicated that a drastic quantum property could be transferred to the seemingly classical world of macroscopic objects. Bohr's insistence on keeping quantum and classical physics apart was much reinforced by this example.

The Refuge in Consciousness

51. I mentioned the idea that the wave function is not an objective physical quantity but a compendium of information: an expression and a summary of the available information for an observer. It is extremely difficult to express it in a sensible way, but von Neumann resorted to it for escaping from his infinite chain. The dull uniformity of an apparatus measuring another, which is measured, and so on, can be broken if one of the apparatuses, the last one, is replaced by an observer endowed with consciousness. This might solve the problem because we know from introspection that an individual's consciousness cannot be multiple, or at least so it was claimed.

It is difficult to understand how this proposal, which was later developed by London and Bauer (1939) and upheld by Wigner (1967), could lure such clever people. The explanation is probably to be found in the cultural tradition of German philosophy, where one finds, for instance, the doctrine of correspondence between physical reality and human consciousness, which had been postulated some time earlier by Ernst Mach and Richard Avenarius. The proposal seems incredible today, not only because cognition sciences have begun to unravel the working of the brain, but also because experimental data are now most frequently "read" by computers and not by human beings.

I feel compelled to add here an aside. The story that great physicists would think of the human mind as the final actor in a measurement, as the deus ex machina transforming indeterminacy into reality, did not remain confined to the physics community. It was great news for the believers in parapsychology and all sorts of occultism: mind acts on matter! This is why, with all the respect one may have and should have for the people who went that far in their hypotheses, one's duty is to emphasize that there is no hint anywhere of its validity *and one does not need it*, as will be shown in the next chapter. The story shows, nonetheless, how difficult the problems of interpretation were.

Schrödinger's Cat

52. The famous article by Schrödinger (1935), which pointed out the difficulties of measurement theory to everyone, was a direct consequence of von Neumann's model. Schrödinger relied on the same example, except that the observable A describes whether a radioactive source has decayed or not. The state 1 in equation (6.1) stands for an intact source and the state 2 for a source where a decay has taken place. The measuring apparatus is no more a simple pointer but a "devilish" device: a Geiger counter reacting to the decay by letting a hammer knock a poison phial. A cat, who is confined with the phial in a box, can be killed by the poison. Despite its vivid character, this is von Neumann's model where the two positions of the pointer are replaced by a cat who can be dead or alive.

Schrödinger also alluded to von Neumann's interpretation when he accepted the linear superposition of the two states of the cat, who is after all a macroscopic measurement device. The conscious human observer enters as the person opening the box. Before that, there is a superposed state (6.2) of a dead and a live cat. When the observer opens the box and sees the outcome of the drama, only one of the two states remains: the cat is either dead or alive.

Although von Neumann's act of consciousness has some similarities with Bohr's reduction of the wave function, there are nevertheless significant differences. According to Bohr, the cat is macroscopic and should have been thought of in classical terms. It has no analogy with a quantum pointer involving only one degree of freedom. Heisenberg had also called attention to the error of assimilating an ideal pointer with a measuring apparatus, in which the number of degrees of freedom is typically of the order of a billion of billions of billions. For both Copenhagen champions, wave function reduction occurred when the Geiger counter reacted, objectively, long before any observer peeped into the box.

Von Neumann's model nevertheless pointed out an essential problem: when the Copenhagen interpretation demands a strict classical description of a measuring device, it obscures a drastic difference between classical and quantum physics. One may then legitimately suspect a logical inconsistency between the classical description of an apparatus and the quantum behavior of its atoms.

It seems that Bohr's and Heisenberg's views on interpretation evolved slightly differently later on. Bohr retained his initial ideas of a classically perfect logic of truth. Heisenberg hinted on various occasions about the effects, not yet called *decoherence*, which could result from the very high number of degrees of freedom in a real measuring apparatus. He did not spurn the possibility of a synthetic formulation of the laws of physics on the unique basis of pure quantum laws, an idea that was slowly progressing.

PILOT WAVES

53. For some people, such as Einstein and de Broglie, the most irksome aspect of the Copenhagen interpretation was complementarity, because it replaced the obvious clarity instinctively attributed to reality by an irreducible ambiguity. De Broglie, then David Bohm and Jean-Pierre Vigier, tried to reformulate quantum mechanics so that the particles really do exist with a well-defined position at every instant of time. They have a trajectory and therefore a velocity. The wave function enters only into the definition of velocity, which is essentially given by the probability current we met in Section 33, $v = \mathrm{Im}(\psi^* \, \partial\psi/\partial x)/\psi^*\psi$ (Bohm, 1952).

When using the Schrödinger equation for the wave function, one finds that the particle motion obeys a Newtonian equation with forces deriving from a potential. This potential comprises two parts: an ordinary one consisting of external actions such as the electrostatic interaction with other particles or an external potential, and also a peculiar quantum potential depending on the wave function. In the case of a unique particle, the quantum potential is given by $Vquant = -(\hbar^2/2m)\,\Delta R/R$, with $R = \psi^*\psi$.

One thus obtains a mixed particle-field theory, where the field is the wave function. These two components of physical reality are not, however, coupled in a symmetric way. The wave function enters into the particle motion, but the "real" position of the particles does not influence the wave function. There is an action and no reaction. One might say that nothing is changed in quantum mechanics but another theory is appended to it for the sole purpose of interpretation. The wave function has become a *pilot wave*.

The idea was brilliant and fascinated some people. It seemed able to describe nonrelativistic physics with an overall agreement with experiments. If a probabilistic distribution for a system of particles at some initial time agrees with Born's rule (i.e., is equal to $\psi^*\psi$), the agreement remains at later times and this is the main positive result of the approach.

There are, however, dire difficulties with relativistic effects that are still not solved after almost fifty years of investigation (Bohm and Hiley, 1993), whereas ordinary quantum mechanics needed less than two years to make the transition to quantum field theory. The problems begin with spin. Is it a real rotation? If so, it would imply difficulties with relativity. The less artificial, "realistic" representation of spin consists in leaving it entirely to the wave function (which would have two components for a nonrelativistic electron), so that the real particle really has no spin. The main problem with relativistic particles is to decide first what is real: the particles or the fields. The ghost of complementarity, which was believed to be killed, is popping up again.

Quantum field theory, particularly electrodynamics, remains outside the range of this realistic interpretation. But this is precisely where one would expect something new from realism, for instance, explaining what the so-called virtual processes mean. Are they more than a picturesque description of a calculation or is there some reality in them?

This aspect of realism is particularly tricky and one can take as an example the vacuum polarization effects in quantum electrodynamics. They can be described with the help of Feynman graphs as the spontaneous creation and annihilation of electron-positron pairs that last a short time, during which they can influence an external field. These effects generate corrections to the Coulomb interaction, and they have been computed with great accuracy. They contribute to important physical effects such as the Lamb shift (essentially a very precise evaluation of hydrogen levels) and the anomalous magnetic moments of the electron and muon. There is excellent agreement between theory and measurements so that the vacuum polarization effects do exist. But in what sense?

In ordinary quantum mechanics, the creation and annihilation of pairs can be considered as a metaphor to help our understanding of dull computations or giving a content to a Feynman graph. This is at least what follows from the consistent histories interpretation to be described later: virtual processes do not enter into "consistent" histories and, more generally, virtual processes do not belong to interpretation. It seems difficult to expect a similar negative statement from Bohm's theory, in which everything is supposed to be real. In any case, I will not discuss Bohm's proposal in more detail because it is far from covering the whole field of physics to which quantum mechanics applies.

EINSTEIN AND INTERPRETATION

54. One cannot say that Einstein was directly interested in the interpretation of quantum mechanics. He believed that the theory contains a large amount of truth but that it remained incomplete. This standpoint was certainly legitimate in 1927, at the time of his great discussions with Bohr, but it lost much of its force when new results and many remarkable applications accumulated with no hint of the supposed incompleteness.

Why did Einstein remain so skeptical? In spite of his famous outburst on God playing no dice, he was not so much dissatisfied by the existence of randomness than by the vanishing character of reality. His thinking about physical reality, whether it be space, time, light or atoms, had been his guide all through his life, with wonderful success. To see a large part of that framework eroded by complementarity was more than Einstein could accept. Yet he never tried to propose a realistic interpretation, as de Broglie and Bohm did, and he was not much interested in such attempts. This is not surprising because the main trend of his mind never brought him to propose a theoretical construction without a preliminary careful analysis of the foundations of a question accounting for the experimental data which are available or conceivable. The paper he published with Boris Podolsky and Nathan Rosen (1935) is an example of his method.

Without yet entering into its details, I will stress what makes the paper novel. For the first time in the history of science and of the philosophy of knowledge, an operational definition of reality was proposed. Although they did not try to define reality completely, Einstein and his collaborators considered that an element of reality exists in a specific case: When "without in any way disturbing a system, we can predict with certainty (i.e., with probability equal to unity), the value of a physical quantity. Then there exists an element of physical reality corresponding to this physical quantity."

They proposed an experimental example where such an element of reality must exist (a better example was proposed later by Bohm [1951]). This existence is claimed to contradict the limitations of knowledge arising from the uncertainty relations. The authors concluded from their example that quantum theory, since it cannot account for the existence of the elements of reality, must be incomplete. This conclusion was criticized by Bohr, and the question to decide whether or not one can define some part of reality remained open.

One of the last statements by Einstein on these questions (Schilpp, 1949) seems to indicate some evolution in his thought. He said about quantum mechanics: "This theory is until now the only one which unites the corpuscular and ondulatory character of matter in a logically satisfactory fashion; and the (testable) relations, which are contained in it, are, within the natural limits fixed by the indeterminacy relations *complete*.[1] The formal relations, which are given in this theory—i.e., its entire mathematical formalism—will probably have to be contained, in the form of logical inferences, in every useful future theory." I will show in the next chapter how this prediction was to be fulfilled with no change in the status of reality.

BELL'S INEQUALITIES

55. Another important event in the history of interpretation was the proposal by John Bell (1964) of an experimental test for the existence of hidden variables. The notion of a hidden variable was contained in de Broglie and Bohm's theory as a "real" coordinate of a particle. More generally, a parameter describing a "real" entity hidden behind the appearances of the quantum laws would be considered as a hidden variable. The existence of this kind of parameter seems a requirement for most scientists who cling to the belief in the existence of a well-defined microscopic reality (contrary to Bohr, who believed in the existence of reality, although not in its sharply outlined characters, which are blurred by complementarity).

Von Neumann believed he could prove that hidden variables are impossible, but there was an error in his proof, which was first noticed by Magda Mugur-Schächter and later by Bell. Nothing in the increasingly precise experiments of particle physics indicated the existence of such parameters. Can there be therefore an experimental test for their existence, even though they are not measured directly? This is the problem Bell considered.

It seems proper to consider a situation very different from ordinary reality, and one is thus led to consider the entangled states of quantum physics. Such a state always involves several degrees of freedom, such as the positions or spins of several particles. Their total wave function is not simply a product of wave functions for the constituent particles but also introduces strong correlations between them.

One may notice that the existence of entangled states is a direct consequence of the symmetry principles for indistinguishable particles. These principles have innumerable consequences which have been tested experimentally, and Bell's approach was not to deny the existence

[1] Underlined in the text.

of entangled states but to turn them to his advantage. He looked at the correlations between the particles in such a state when they are very distant from each other.

One knows that long-distance correlations exist in classical physics. For instance, the shapes of the two halves of a torn up bill remain correlated, as has often been seen in detective stories. Bell gives as another example the socks of Mr. Bertlman. One sock is always green and the other is red. When you see one sock, you know the color of the other. This is an obvious case of correlation. In quantum mechanics, the color of one sock can be replaced by some eigenvalue for an observable associated with a definite particle (which is one of Bertlman's feet). There are cases where this observable is correlated with another one belonging to a different particle.

The essential difference with the classical case is that the correlation can persist when the observables are changed. This can be true for the spin component of a particle along a definite direction, in which case the spin component of the other particle along the same direction can show a correlation, however large the distance is between the two particles. There are, in other words, some cases when a large distance is not enough for separating the properties of two objects; they are then said to constitute a nonseparable system. To see the difference with the classical case, one might pursue Bell's example and replace the particles having spin with two socks. The two socks are apparently white to the naked eye, but they show complementary colors when one looks at them through complementary color filters, for instance a red or a green filter. One sees a sock to be red. The correlation implies that the other sock will be seen to be green. But if one uses a pair of yellow and purple filters (similarly to a spin measurement along another direction), the correlation implies that one sock is yellow when the other is purple. This game with filters has certainly no analogue in classical physics.

If hidden variables exist, Bell assumed them to be separable. This means that choosing a direction for measuring the spin of one particle on Earth, for instance, does not introduce a correlation between the local hidden variables and other hidden variables for another apparatus and another particle in the Andromeda galaxy. This separability assumption is supposed to be characteristic of commonsense realism, whereas quantum mechanics has no such constraint.

Bell's brilliant idea was to find an experimental setup where the difference between the assumptions of separable realism and quantum mechanics result in opposite conclusions. This will be discussed in Chapter 22, where some of the experiments that have decided in favor of quantum mechanics are described (Aspect, 1981, 1982). One must acknowledge for the sake of rigor, however, that some far-fetched possibilities for hidden separable variables remain, although they would be quite strange from the standpoint of commonsense realism. It may also be noticed that the models with a pilot wave are not refuted by these results because their hidden parameters can be correlated in a nonseparable manner through the wave function.

SOME OTHER FEATURES OF INTERPRETATION

56. All things considered, after the formulation of the Copenhagen interpretation, more than fifty years elapsed without an essential change. This rather quiescent period was followed recently by an intensive series of experiments and new theoretical works, which will be described in the next chapter.

Before entering this new episode, I will mention briefly a few significant results that occurred in the interim. One of the first interesting results was obtained by Nevill Mott (1929a), who wanted to understand why a particle track in a detector (a Wilson chamber, for instance) can be a straight line even when there is an isotropic spreading of the wave function. Mott considered the interaction of an electron with several atoms, which can become ionized in a collision. He computed the joint probability for several ionization events and found that it is negligible when the atoms do not stand along a straight line. This is not very different from the emergence of geometrical optics from wave optics. I will have other occasions to come back to this point.

Some limitations on the possibility of a quantum measurement were also realized. The first example came from Bohr, who considered the constraints of complementarity on the measurement of spin and concluded that it is impossible, as a matter of principle, to measure a spin component of a free electron. The reason is that the charge of the electron and the value of its magnetic moment imply an interference effect between the Lorentz force and the magnetic force in a Stern-Gerlach device: the two forces do not commute. (The argument does not apply to a bound electron in a hydrogenoid atom because of the atom neutrality and the absence of a Lorentz force.) Bohr's argument was worked out by Mott (1929b) with all the necessary calculations.

In a similar vein, Wigner (1952), and later Huzihiro Araki and Michael Yanabe (1960) found that an observable that does not commute with the constants of motions cannot be measured exactly in an ideal von Neumann measurement. It can, however, always be measured with an a priori *given* precision, at least in principle, so that the theorem is of little practical avail.

On the other hand, Yakir Aharonov and David Bohm (1961) have shown that a quantum system can be sensitive to some observables that cannot be measured in classical physics. A magnetic vector potential, for instance, cannot be measured classically. One can, however, observe quantum interferences when an electron can go around a solenoid, which generates a vector potential outside but no magnetic field. The phase difference between the two parts of the wave function passing on both sides of the solenoid involves the well-known relation $p' = p - eA$ between the kinetic momentum p' of the electron and the Lagrange momentum p, which is canonically conjugate to position. The Aharonov-Bohm effect was observed for the first time by Jaklevic et al. (1964).

It should be mentioned finally that the older literature on interpretation considered that a measurement must necessarily disturb a measured microscopic object. This argument was often used for an empirical justification of complementarity. The argument was wrong or at least not universal, as shown by Braginsky, Vorontsov, and Thorne with the existence of nondemolition measurements (1980).

7

The Present State of Interpretation

57. Perspective is lacking for objectively appreciating the most recent tendencies of interpretation. There is nevertheless a significant renewal in both experiments and theory, particularly since the 1980s. Although this date can only be indicative, it marks a transition between a period when Bell's ideas and the questions concerning hidden variables were dominant and a return to the origins—the basic problems first envisioned by Bohr, Heisenberg, and Pauli. The new trend is to readdress the interpretation of standard quantum mechanics without the use of incomplete and questionable exotic theories.

There is a renewal in experiment. Many experiments that were previously impossible can now be performed with modern techniques. One can thus continuously observe a unique atom or build up macroscopic systems that behave in a quantum way. Delayed-choice experiments have been made, in which the decision of the measurement to be made is taken when a photon wave function is already inside the two arms of an interferometer. It looks almost as if everything is becoming possible.

The renewal of theoretical ideas is certainly as important. I will particularly highlight three of them: the decoherence effect, the emergence of classical physics from quantum theory, and the constitution of a universal language of interpretation by means of consistent histories. These advances will be described briefly, at the same level of generality as in the two previous chapters, before developing them in detail in Part Three. As an initial orientation, one can say that the decoherence effect, which was recently observed, can explain the absence of macroscopic interferences and solve the Schrödinger's cat problem. The second development is a strong improvement in understanding the relation between quantum and classical physics, which is now well understood and succeeds in reconciling determinism with probabilism. Finally, the method of consistent histories provides a logical structure for quantum mechanics and classical physics as well, that is, the universal language of interpretation initially expected by Bohr, Heisenberg, and Pauli.

When these three ideas are put together, they provide a genuine theory of interpretation in which everything is derived directly from the basic principles alone and the rules of measurement theory become so many theorems. As we might expect, a few points remain and new problems arise, but they are by far more technical and less essential than the ones we met previously. In any case, let me be more explicit.

COLLECTIVE OBSERVABLES AND ENVIRONMENT

58. I will begin with one of the points that are not yet fully worked out. The question at hand is to distinguish what is directly accessible to observation in a macroscopic object from what is not. I will briefly review the various approaches that have been used or attempted for answering this question, although the one we will finally adopt is very simple and pragmatic. There is therefore no inconvenience to the reader if he or she skips the following discussion (in smaller type).

The problem is concerned with the definition of *collective variables* (or *observables*) and it already exists in classical physics when one introduces general coordinates in the Lagrange-Hamilton scheme. Nowhere is there a general rule for choosing, or even guessing, the right coordinates. Who will tell us, in a complex case, that we have the right variables and that the ones we chose are sufficient for our purpose? The situation is the same in thermodynamics, and finding the relevant variables is already a significant step in the understanding of a physical situation. This is the well-known problem of "putting a system in equations," which begins by identifying the relevant variables. The work involved is mostly a matter of unwritten expertise, and it requires a large amount of creativity when one is dealing with a new or unusual case.

One often proceeds by trial and error. For instance, a car is considered first as a solid frame undergoing horizontal motion, the wheels being acted on by a motor while having friction with the ground. This model can be used in undergraduate exercises but is very far yet from the details a specialist must take into account: the elastic deformation of tires (and whether they are elastic), the moving parts in the engine, vibrations, the working of electric and magnetic appliances and—why not?—the motion of fluids and the working of a control microprocessor. When does one stop? The answer is usually two-fold: when calculations become too complicated and/or when some estimates of order of magnitude show that something can be neglected, *at a sensible level of precision*. One can never do better when dealing with a real physical system, except in an excessively simple case. The idea of considering all the atoms in a car, even if their interactions and their motion were very simple, is obviously clumsy and useless. The fact of knowing how to write out the quantum Hamiltonian for all these atoms and particles is certainly not of much help.

In ordinary practice, the Hilbert space and the Hamiltonian for all the constituent particles remain mostly a dream in our mind. They are much too general when one realizes that the particles in the car, with the same Hilbert space and the same Hamiltonian, could as well make two or three motorcycles or (allowing a redistribution of the nucleons among nuclei) the sweet-smelling content of a flower-shop.

Perhaps the fuzziest idea in quantum mechanics is the concept of "object." In a simple example, like a hydrogen atom made of only one electron and one proton, one can get a hint of what "object" means. The atom is a bound state, as distinct from two separate particles. It is therefore associated with the Hilbert subspace that is spanned by all the bound states. One may think that, analogously, every well-defined object is associated with some Hilbert subspace, but this idea remains rather vague. How are we to define it precisely? What happens when a further atom is added to the object? Is it possible to define some observables as properly belonging to a well-defined object?

59. The question of defining *relevant observables* has another aspect in statistical mechanics, where the thermodynamical limit is obtained by letting the numbers of degrees of freedom go to infinity. One can then define a collective observable by a mathematical procedure: it must commute with all other observables (as a matter of fact, the Hilbert space framework does not fit a system with an infinite number of degrees of freedom and one must use C*-algebras). The collective observables are then automatically classical because they commute together. This approach has sometimes been considered essential for introducing classical physics into interpretation (Primas, 1981). The proposal is doubtful, however, because a real object is never really infinite and there are then very few observables commuting with all others, if any.

Another approach is coarse graining, which has been proposed particularly by Gell-Mann and Hartle (1991, 1993). It consists in restricting attention to regions in configuration space (or phase space) that are both microscopically big and macroscopically small. The relevant observables are then obtained by an average on these regions of some well-chosen particle observables—number and position, for instance. There is also the chemists' approach: The overall shape of a molecule is considered as fixed and its motion as classical. Vibrations are described by the internal position of nuclei, which can be treated quantum-mechanically by using the Born-Oppenheimer approximation. Other methods for introducing relevant observables are used in nuclear physics, for instance, when dealing with highly deformed nuclei. I cannot review all the methods, but in any case, one is very far from the dream of starting from scratch in a vast

Hilbert space, having a criterion defining the various objects it can represent and an algorithm producing the relevant observables (see, however, the last pages in the article by Charles Feffermann [1983] for possible mathematical hints).

Finally, from a fundamental standpoint, it is impossible to define exactly the concept of object and to define from scratch a set of relevant observables. Fortunately, this is mainly a problem for a purist and has no direct bearing on the basics of interpretation.

As a matter of fact, one knows what to do and to think when dealing with a real object under one's eyes. Somebody has practically always been able—sometimes obviously, or by guesswork, or with a touch of genius—to find good, relevant observables for describing an object and even a hierarchy of observables according to their relative importance in dynamics. When this has been done classically, it is easy to express those observables in a quantum framework. This method, which consists in considering as obvious what looks obvious enough, will be sufficient for our purpose.

60. I will mainly rely on a pragmatic conception of a macroscopic object. Its overall dynamics is supposed to be described, at a sufficient level of precision, by a set of relevant observables. They are often called *collective observables*, although this may be slightly less general than *relevant observables*. The position of the center of mass of a golf ball, for instance, is typically a collective observable (it is easily expressed in terms of the position observables of the constituent particles). On the other hand, the momentum of an electron in a Geiger counter can be relevant for the working of the counter although it is not collective. One may also notice that relevant observables are not necessarily defined once and for all. When, for instance, a charged particle produces a track of bubbles in a bubble chamber, the position of each new bubble appearing is a new relevant observable.

In a given situation, one can describe a macroscopic system by using a complete set of commuting observables, some of them acting as relevant coordinates Q and the rest of them being microscopic coordinates q. The latter can describe the internal constituents of the object (nuclei and electrons), as well as its external environment if necessary (the air molecules around the object or the photons in ambient light). The total wave function is then a function $\psi(Q, q)$.

The domain to which the relevant coordinates belong is a configuration space, which is generally not a Euclidean space. This is a consequence of a deep theorem by Gelf'and and Segal about commutative algebras (Naimark, 1959), but it is also familiar in ordinary practice, and this will be sufficient for our purposes. There is also a natural measure on configuration space allowing one to define a scalar product of wave functions and a collective Hilbert space $\mathcal{H}_c$, although I will not discuss it here. There is also a Hilbert space $\mathcal{H}_e$ associated with the q variables, and the total Hilbert space is the (tensor) product of these two.

Although abstract, this approach means that one can always think of the object as split into two subsystems, one of them being a "collective" subsystem described by the Q's and the other, described by the q's, being called the *environment*. This name suggests an atmospheric or photonic environment surrounding the object, but it has been conventionally extended to anything the variables q can describe and therefore, in some way, all the internal details of matter and radiation in the object and outside it.

Most existing theoretical discussions are devoted to the rather frequent case when the total Hamiltonian of the object can be written as a sum

$$H = H_c + H_e + H_1, \tag{7.1}$$

where H_c is an operator acting in the relevant Hilbert space, H_e is an environment operator, and H_1 a coupling term. These notions look familiar if one thinks of thermodynamics. H_c is the energy, kinetic and potential, for the collective motion one can see at a large scale. H_e is the internal energy and, more generally, represents an internal motion (mostly thermal). As for H_1, it is a coupling producing an energy exchange between the two kinds of motion through dissipation (Joule effect, friction, and so on).

It may be noticed that equation (7.1) is too simple in many realistic situations. The case of friction between two solid objects requires, for instance, a more refined analysis (Caroli and Nozières, 1995; Tanguy and Nozières, 1996). One may think also of the intricacies of fluid friction, but in any case, one cannot expect to include within a unique model the various effects that can be met in practice. The model described is general enough for allowing a reasonable investigation of the most important quantum effects, and it is used most often as a reference and an example.

DECOHERENCE

61. The most worrying difficulty in the interpretation of quantum mechanics is certainly the problem of macroscopic interferences, which are apparently predicted by any linear theory and practically never observed, so much so that they would look absurd if we were to see

them. Reflection on this problem led to the idea of decoherence, which is certainly the most important discovery of the modern interpretation.

It will be convenient to go back to equation (6.2), which expresses the superposed state of a measuring device after a measurement. I will assume for simplicity that the measured system has been destroyed during the measurement. I will consider the final wave function for the apparatus depending on the position variable x of a pointer as given by

$$\psi(x) = c_1\phi_1(x) + c_2\phi_2(x), \tag{7.2}$$

where $\phi_1(x)$ is a wave function centered at a point x_1 and $\phi_2(x)$ is centered at a point x_2 macroscopically distant from x_1. All the problems discussed by von Neumann and Schrödinger are contained in this simple expression.

If one thinks of a realistic (macroscopic) measuring device, the position x of a pointer is a collective coordinate. There are many other coordinates, which can be denoted by y and which describe the environment, so that rather than equation (7.2), one should write more explicitly

$$\psi(x,y) = c_1\phi_1(x,y) + c_2\phi_2(x,y). \tag{7.3}$$

Heisenberg, on various occasions, had called attention on the possible role of the environment in the problem of interferences. But what is it exactly? The answer was given by Nicolaas van Kampen (1954) and was later rediscovered by Hans Dieter Zeh (1970). When one tries to guess how a function such as $\phi_1(x,y)$ may depend on y, one expects it to be a very complicated function, which means a function with a very rapidly changing phase for slightly different values of y. The motion of a pointer as it occurs before reaching a final position indicating a datum (or any other macroscopic motion) must be a cataclysm in the little world of the y's. Many atoms in neighboring pieces of the apparatus rub against each other, and this entails certainly drastic changes in the state of the environment which cannot be seen macroscopically but which affect strongly the mathematical details of the wave function. One may thus expect that the two final functions, $\phi_1(x,y)$ and $\phi_2(x,y)$ are very different in their fine y dependence. This means particularly that for the same value of y they have very different phases, which vary strongly in each wave function with no direct relation to each other.

The outcome of such a complete lack of phase coherence cannot be other than orthogonality of the environment part of the two wave functions, that is,

$$\int \phi_1^*(x,y)\phi_2(x',y)\,dy = 0, \tag{7.4}$$

for any pair of values x and x'. This orthogonality property implies the suppression of macroscopic interferences when only the x-variables are observed.

62. If there is really a suppression of macroscopic interferences along these lines, a sensible description of a measurement must be somewhat more sophisticated than the simple von Neumann model. One must take into account the interaction of the collective degrees of freedom with the environment. Furthermore, because time evolution is continuous, the loss of phase coherence in the functions of y must evolve gradually and equation (2.4) is only a final outcome. Decoherence, or the absence of macroscopic interferences, is therefore not a given intrinsic property of nature but rather a dynamical effect requiring some time to become effective. But how much time?

This problem was first investigated with the help of simple models for the environment and then under more general conditions until reasonable conclusions were reached for the main questions. Research is still continuing actively on this topic and there may soon be a universal understanding of the decoherence process. Still, the results already obtained are very instructive. A very short time t_d characterizes decoherence, and the integral in equation (7.4), which can be used as a measure of the effect, decreases exponentially as $\exp(-t/t_d)$. The effect is very efficient, and a few atmospheric molecules or a few photons hitting an apparatus can produce significant decoherence. Or it can be caused by a few phonons that are produced through friction between different parts of the apparatus, or by a few electrons that start an electric current in the electronic components. Wave functions of a macroscopic object are extremely sensitive and macroscopic interferences are therefore never observed except under very special conditions, which will be considered below.

Decoherence is certainly the most efficient and rapid effect to exist in macroscopic physics. When this simple and basic fact was recognized, one had at last the key to the cat problem. There is, however, some sort of a paradox in the effect: it is so efficient that one does not see it in action. It has already been completed before an experimental device could catch it. Can one then see the decoherence effect in action? It is clear that if one could make a measurement with an apparatus involving only a few environment variables and if one could vary the number of variables, the effect would be detectable. One would see quantum interferences, and then see them vanish gradually with an increasing number of parameters. This was essentially done in an experiment led by Jean-Michel Raimond and Serge Haroche (Brune et al., 1996), which is described in Chapter 22. Its conclusion is clear and simple: decoherence exists and observations of it are in excellent agreement with theory.

63. I said that macroscopic quantum interferences are never observed, but this was an imprecise statement and the correct conclusions are more subtle. Decoherence depends on the coupling H_1 in equation (7.1) between the collective phenomena and the environment. If this coupling happens to vanish, there will be no effect of the environment and quantum interferences will remain. Is this possible? I already mentioned that H_1 is also responsible for dissipation effects, and as a matter of fact, this common origin of decoherence and dissipation is reflected by a proportionality between the times over which they act. The elusive decoherence time is much shorter than the time necessary for dissipative damping, but one can nevertheless assert that a macroscopic system with no dissipation has no decoherence.

The most obvious nondissipative macroscopic system is light, which is macroscopic when it involves a large number of photons. Though not strictly zero, the interaction between two photons is extremely small, and there is no sizeable dissipation in the case of light. This is why one observes macroscopic interferences with ordinary light, as well known. Superconductors provide another example of a nondissipating system, and clever devices using them have also shown purely quantum effects at a macroscopic level.

THE CORRESPONDENCE "PRINCIPLE"

64. I will now turn to the second problem, which is the correspondence between classical and quantum physics. Some collective features of a macroscopic system are clearly seen to behave classically. This evidence, however, does not explain why the collective phenomena happen to be classical, granting that the quantum laws of physics are universal. The fuzzy correspondence principle dating from the origin of quantum theory is not enough. One needs quantitative correspondence theorems.

The analogy between the transition from quantum to classical physics and from the wave theory of light to geometrical optics is well known. This relation was most helpful when Schrödinger "quantized" classical physics to develop his famous equation. The real problem now is exactly the inverse: to understand classical physics on the basis of quantum mechanics. It is obviously not easy and raises many questions. How can one switch from quantum observables to the ordinary dynamical variables of classical physics? What is the meaning, in a quantum framework, of a classical property in which one asserts simultaneously the values of position and momentum? How can one derive the equations of classical dynamics from the Schrödinger equation and with what kind of errors? How can one reconcile classical determinism with quantum randomness? And since determinism does not apply in a chaotic system, what does the correspondence principle say in that case? Finally, during the early history of quantum theory, the abstract Hamilton form of classical dynamics took precedence over the intuitive description by Newton. The Schrödinger wave functions were then found to exist in an abstract configuration space and not in ordinary three-dimensional space. So a last question would be: how can one make the final step backward and explain from quantum mechanics why we see the world in a three-dimensional space? These questions, which look so strange from the viewpoint of common sense, must be answered when classical physics is considered as a consequence of more fundamental laws.

They have recently been answered. The resulting emergence of classical physics from a quantum substratum is now essentially complete. The relation between determinism and probabilism is the easiest one to explain in a few words and I already mentioned it: determinism is valid, *up to some probabilities of error, which are extremely small in most practical circumstances*.

The recourse to the numerical values of some probabilities for finding a way out of an apparently strong contradiction shows that the motto should be: be quantitative! This practice works, but it implies a price to be paid: the problems to be solved involve nontrivial theoretical developments because otherwise the answers would have been known much earlier. As a matter of fact, their answer relies on rather recent results in mathematics that I will not discuss in this chapter. I will therefore summarize them in a sweeping statement: Yes, all the features of classical physics derive directly from quantum mechanics. Most macroscopic systems behave classically, although there are exceptions that are well under the control of theory and in agreement with observation.

HISTORIES

65. I said earlier that von Neumann was able to translate a sentence expressing a quantum property (predicate) into a mathematical notion: a *projection operator*, which I shall henceforth call a *projector* for short. Von Neumann had stumbled, however, onto two obstacles: the extension of his method to classical phenomena and dire difficulties with logic. These two problems have since been solved through the approach of consistent histories.

There is still some confusion about the status of consistent histories, which is why, before telling what a history is, I will clarify the sense of the approach itself. I believe the best way of expressing it is to start at the very beginning of interpretation, when Bohr, Heisenberg, and Pauli were in search of a universal language for interpretation, before complementarity had taken precedence. Histories can be considered as that language. They are able to describe anything existing or happening in a definite physical situation at the microscopic (quantum), as well as at the macroscopic (classical), level. Such a description is the primary role of a language. The present one provides also a well-defined logical frame, which is necessary in a scientific language. Complementarity is one of its consequences and certainly not a substitute for it, as Bohr supposed. When seen as a language, histories, furthermore, show no ambiguity in their relation with reality. A language can describe everything that is reliable but also possible and even conceivable; it can therefore recount real events if necessary, but it does not decide what is or should be real. The aim of histories is a clear organization of its subject matter—interpretation—in a methodical way.

But then, what is a history? Von Neumann had made clear that the language of projectors can describe any quantum property. But a description of what happens during an experiment does not consist of a single property. Many properties must be used, like a story, which needs many sentences for being recounted and like a motion picture, which consists of many successive photographs. *A history is accordingly a sequence of properties (predicates) referring to well-defined successive moments of time and describing the events occurring in an experiment.*

A history will say how a quantum system is prepared, what kind of apparatuses take part in the experiment, how the experimental devices work, what happens at the level of atoms and particles, and which data come out of a measurement. This extraordinary versatility is due to an essential advance over von Neumann's initial language. Whereas he could account only for the properties of atoms and particles, the recent powerful techniques yielding classical physics can considerably extend the range of the construction. Classically meaningful properties of the preparation device and of the experimental apparatus—what we can see and do, phenomena, and actions—can now be expressed in the language of quantum projectors. This is why this language can now be considered as universal.

One may ask what became of the difficulties with logic that seemed to preclude a meaning for such a language. They were disposed of in two steps. Robert Griffiths (1984) made the most significant advance by noticing that many histories can be put forward with the intent of describing physical events but that many of them are nonsense. The standard example is an interference experiment in which one tries to say through which hole in a Young device a particle has gone. This is nonsense. Some histories, however, or rather some families (or classes) of histories listing mutually exclusive events, can make genuine sense. They are recognized from the fact that a perfectly valid probability exists for each history in the family. The main criterion for the existence of these *consistent histories* is the consistency of additive probabilities with additive quantum amplitudes. This criterion is expressed by a set of consistency conditions which select meaningful histories from the senseless ones. These conditions are explicit equations involving the projectors for the various individual properties (predicates) occurring in the histories of the selected family.

The relation of probabilities with logic was the second step, and it was clarified by this author. It was shown that *ordinary logic holds in a family of consistent histories*. The standard rules of logic (the logical operations *and*, *or*, *not*, as well as inference and logical equivalence) rely on the existence of these probabilities (Omnès, 1988). This logic is formal, but is not quantum mechanics itself a formal theory? What is most important when one remembers von Neumann's difficulties is that the present logic is perfectly conventional. One might call it Boolean, or Aristotelian, or whatever name one wants to give it, but in any case it is very different from the exotic kinds of logic that had been envisioned by Birkhoff and von Neumann. It is the standard kind of logic that has been tested by the construction of mathematics and by sound reasoning in the field of natural sciences.

I will show in later chapters how complementarity can thereby be defined explicitly. It means essentially that *several* consistent families of histories can describe the same physical system and the events occurring in it. Each family is logically consistent internally but two of them are not necessarily consistent with each other. This mutual exclusion is the essence of complementarity, which is therefore a feature of the language of interpretation and the reason why there is no unique description of microscopic reality.

The consistency conditions were related to decoherence by Gell-Mann and Hartle (1991). As a matter of fact, logical consistency is most often a consequence of two basic features of events (or more properly of either one of these features or of both of them): (1) when classical physics holds, as when thinking classically of a particle in an accelerator or when a particle moves from one piece of apparatus to another; (2) when decoherence acts. This occurs mainly when a measurement takes place.

Some readers might wonder in view of this sketchy description whether the language of histories is not too abstract and too far from empirical physics. The opposite is true because the language is in fact more or less trivial. It is the everyday language one hears physicists using when they describe the events in an experiment, whether it be a quantum experiment or not. Its logic is mostly common sense, when due caution is taken concerning complementarity.

A DEDUCTIVE INTERPRETATION

The Classical Framework

66. The close relationship between classical physics, decoherence, and the language of consistent histories may suggest that a complete interpretation of quantum mechanics is close at hand. If so, one should be able to make sure by constructing a completely deductive interpretation. The only hypothesis would be the assumption of the basic principles of quantum mechanics. Every step in the interpretation would be secured by a proof and, when possible, by experiments. One should also make sure that no loopholes exist. Although this is a rather ideal description of a theory of interpretation, it is essentially what has been obtained recently (Omnès, 1990, 1992) with satisfactory results for the logical consistency of interpretation. Previous attempts in the same direction had been made by Klaus Hepp (Hepp 1972, 1974a; Hepp et al., 1974b) or sketched by Gérard Emch (Emch, 1972; Whitten-Wolfe and Emch, 1976). These authors, however, were not able to reach a final conclusion, because of an insufficient understanding of decoherence at that time and perhaps, because of the lack of a logical guide that the language of consistent histories would later provide.

The new approach sometimes implies a significant change in standpoint, and I will give as an example the correspondence between quantum and classical physics. This is particularly interesting because the three main ingredients of the theory (semiclassical physics, decoherence, and histories) interact to provide a full expression of correspondence.

The classical language, in which position and momentum are mentioned together, is obviously valid empirically when one is describing a macroscopic object. Its agreement with the language of quantum mechanics (or, rather, with von Neumann's projectors) is far less immediate, but it can be proved with good mathematics.

This ordinary language has its own logic, which is essentially common sense. The logic of common sense can be proved to agree with the principles of quantum mechanics within the language of consistent histories and, of course, when dealing with macroscopic objects. Since it mentions only classically meaningful properties, it does not suffer from the ambiguities of complementarity.

One can then show that the Schrödinger equation implies classical dynamics. This is rather easy when friction is neglected; one obtains the Lagrange-Hamilton rules for classical dynamics as a consequence of quantum dynamics, at least for collective variables. There is, of course, some difference between the two kinds of dynamics, and some errors occur when applying the classical one. These errors are due to the finite value of Planck's constant or, if one prefers, to quantum fluctuations. They are negligible in most practical applications and even exponentially negligible. There are, however, some cases when the classical-quantum correspondence is violated, for example, in a chaotic system on the verge of reaching the scale of Planck's constant or when the potential barriers are very narrow. These limitations of correspondence (which stand to reason) are met in the theory as expected exceptions to a few basic theorems, or rather as a special case when the estimate of errors resulting from these theorems are too large.

Determinism can be thereby considered, from the standpoint of logic, as a logical equivalence between two classically meaningful properties occurring at different times. The limitations on the validity of determinism are the same as for dynamics, and the errors are the same. The error in determinism is the probability for one of the classical properties to be false when the other one, to which it would have been equivalent, is supposed to be true.

Decoherence is the last touch in completing the usual representation of the world. As a matter of fact, the Lagrange-Hamilton version of classical physics which I have said to be already obtained, is invariant under canonical transformations. It sees nothing special in the Newtonian representation of physics in physical three-dimensional space, and it concerns only an abstract phase space. But *decoherence has the remarkable property of selecting a special choice of collective coordinates*. In the case of a purely mechanical macroscopic system involving solid or fluid parts but no macroscopic electromagnetic effects, decoherence selects the collective coordinates in three-dimensional space (because of the invariance of the laws of physics under a change of reference frame). Decoherence also goes with friction.

We might express these results by saying that one can completely recover Newton's intuitive conceptions of a classical macroscopic world and show them to be in agreement with a subjacent quantum mechanics.

67.* If we allow ourselves to indulge in more philosophical considerations, some further remarks are perhaps worth mentioning. One can reject Bohr's prohibition on describing the atomic world (except by mathematical means). Such descriptions are, however, very convenient, and their use is almost universal among physicists. They must sometimes be checked, however, and when a doubt arises about their validity in a specific case, Griffiths' consistency conditions must be used as a criterion. Complementarity cannot be avoided in any case, and it maintains a gap between the language of histories and the usual category of reality implying uniqueness. One must get accustomed to this veiled reality, as Bernard d'Espagnat called it (d'Espagnat 1985, 1994), and one certainly cannot expect a return to the full reality of yore, when everything could be seen with human eyes.

MEASUREMENT THEORY

68. I will only briefly consider measurement theory in so far as most of it reduces to a derivation of the old Copenhagen rules, as given in Section 44. A somewhat new feature (see, however, Jauch, 1964) consists in a careful distinction that must be introduced between a quantum property holding at the beginning of the measurement process and the final datum. The quantum property, which will be the *result* of the measurement, is what one wants to learn through the measurement. It can be, for instance, the value of a spin component. The datum is, on the contrary, a classical phenomenon, such as the position of a pointer or a number that is registered in a computer memory. Quantum logic shows that the result and the datum are logically equivalent when the measurement is correct, so that the conflation one usually entertains between the two of them is well justified.

All the rules given in Section 44 become so many theorems. This is particularly true for wave function reduction when it is stated as a convenient recipe for the probabilities of the results in a second measurement when the result of a first one is known. The rule is extracted in the new approach from some consistent histories where the various results and data are taken into account, and in practice, the reduction rule is a summary of these histories. There is no physical reduction effect acting on the measured system and the only effect that must be necessarily at work is decoherence, which acts in measuring devices.

THE PROBLEM OF OBJECTIFICATION

69. Objectification, as we have seen, is a word for a fact. The fact is that one observes a unique datum as the outcome of a measurement. Objectification is still presently a matter of controversy. The uniqueness of the datum seems obvious "because reality cannot be else than

unique," but the theory stubbornly keeps the various possible data on the same footing, and this is again an apparent source of contradiction.

A significant result of the new interpretation addresses this problem: a measurement datum must stand once and for all after its establishment, so that it can be considered as a fact (this is true, by the way, of any macroscopic phenomenon, whether or not it originates in a quantum event). This is to be contrasted with von Neumann's model, where in keeping with the Schrödinger cat analogy, a cat revealed as dead from a measurement could very well be alive later when another measurement is performed on the system to which it belongs. The notion of fact had no sense in pure quantum mechanics before the discovery of the decoherence effect.

When decoherence and the quantum version of classical physics are taken into account, one finds something very different from the older von Neumann–Schrödinger situation. When a history includes a phenomenon that is specified by decoherence, *there can be no consistency for a later property that would contradict this phenomenon or its consequences*. One cannot logically deny it. It gives rise to an indelible record, that retains its consequences, even if it is erased or dissipates. It remains present in the inward details of the wave functions, decoherence forbidding the consistency of its negation. Any history that would try to deny it (or its later consequences) necessarily violates the consistency conditions and therefore the rules of logic.

It is also pertinent to re-emphasize the irreversible character of decoherence, which selects the same direction of time as thermodynamics. Quantum logic has also a definite direction of time, which must be the same as that of decoherence because consistency is most often a consequence of decoherence. The words "later," and "earlier" therefore have a definite meaning.

Another problem brings together the direction of time and objectification. It was raised by Bell (1975) and d'Espagnat (1990, 1994) and is expressed schematically as follows: does the decoherence effect provide a fundamental explanation for the existence of facts, or is it valid only for all practical purposes, "saving appearances" in a questionable way? Bell and d'Espagnat inclined toward the second answer. The experimental observation of decoherence changed the situation and, in my opinion, supports the first answer. Experiments reveal cases where decoherence is partially or fully developed, having a quantitative character just as in the case of classical physics. One must therefore face a larger problem, which is the qualitative meaning of assertions relying on probabilities nearly equal to 1, when these assertions justify common sense (viz., classical physics) or objectification (viz., the status of facts).

The phenomena resulting from decoherence are almost irrevocable, except for incredibly small probabilities of errors, just like classical physics and its determinism. Quantum theory cannot, however, completely remove the errors or the exceptions as a matter of principle. One can always dream of a giant quantum fluctuation by which, for instance, the ink molecules on the pages of the present book would move to turn it into perfect English. This kind of event, like a dead cat returning alive from a subtle measuring operation, does not belong to the empirical physics. On the other hand, when one is envisioning a turn away from decoherence or a tunnel effect which would send the Earth revolving around Sirius, one is introducing the problem of the meaning of extremely small probabilities. What do they mean, and should one consider them as something whose existence is "fundamental"? Or can one say that a too small probability does not make sense from the standpoint of physics?

Currently, there is no general agreement on these questions, and some people consider these problems as "fundamental," whereas others say that there is no problem at all. Bell and d'Espagnat's request for a "fundamental" explanation of objectification is tantamount to the old assertion of incompleteness of quantum mechanics: quantum theory, according to them, would not explain objectification because of nonzero probabilities for overturning decoherence. Their opponents claim that one should not care about these tiny probabilities.

Fundamentalists in a science can also encounter opponents from another more fundamental science. When discussing the meaning of very small probabilities, one should listen to what the specialists have to say. Probability calculus is an example of a formal science needing an interpretation. I only know of one scientist who had a try at this problem, although they may be many. Émile Borel, one of the founders of modern probability calculus, gave much attention to the interpretation of this theory. He was particularly interested in the meaning of very small probabilities and he proposed a unique axiom for interpretation: one should consider that an event with too small a probability will never occur (Borel, 1937). His arguments are detailed and careful, but the main one is that this kind of event cannot be reproducible and should be left out of science. Mathematical theory cannot avoid the consideration of these events, but what it says about them cannot be held reliable nor even sensible. If one accepts this lesson, it means that the explanation of facts resulting from decoherence is really fundamental, discarding objectification as a real problem. Quantum theory would therefore be complete. This will be the main conclusion of the present book.

Should one be more cautious? Of course, this conclusion assumes the *empirical* validity of present quantum mechanics and could be refuted on some remote frontier. After all, we have no idea why this theory holds. "*hypotheses non fingo*" as Newton said, which precludes answering in the present case John Wheeler's question, "why the quantum?" But it works and one can understand it.

PART THREE

Reconstructing Interpretation

This third and largest part of the book describes in more detail the interpretation sketched in the previous chapter. It relies on three main ideas: (1) the language and logic of consistent histories; (2) the correspondence principle converted into a few basic theorems; (3) the decoherence effect.

I will explain most of the theoretical results that will be needed in an elementary way, as far as possible, and will not use higher mathematics. Doing so may hinder rigor, but the requirements for understanding and those of mathematical rigor are not necessarily identical. The construction will proceed systematically, and rather than giving in advance an indication about the content of each chapter, I will conclude each with a summary of results, as a guide in the construction of interpretation.

8

Principles

70. The present chapter will be devoted to the foundations on which to build an interpretation. I must therefore begin by stating the principles of quantum mechanics on which I intend to rely. There is some arbitrariness in their statement because several versions are possible. The choice between matrix mechanics, wave mechanics, or Feynman path integrals is an example. Some foundations may be redundant, as when one principle can be derived from others. But the most economical set of principles is not necessarily the most convenient, and one must make a choice. The one I made does not pretend to be better or worse than any other, but it is deliberately directed toward interpretation.

I will assume only four principles. The first one determines the mathematical framework of the theory. The second and third are the principle of invariance under a change of frame and the law of dynamics (according to Heisenberg and Schrödinger). The last principle is concerned with the symmetry properties of identical particles. In these preliminaries I will say nothing about quantum probabilities nor about the state of a system, including the notion of a wave function, because these questions require some interpretation. They will therefore be postponed temporarily. These principles are not only the achievement of the historical discovery I described earlier. Each of them has been tested in many applications, so that they can stand as a summary of a vast amount of data.

My considerations will not be restricted to quantum principles. I will also rely on other foundations, including those of logic and probability calculus. All the axioms that will be used later are therefore collected in the present chapter.

THE HILBERT SPACE FRAMEWORK

71. We saw earlier how von Neumann was led to propose the framework of Hilbert spaces as most convenient for quantum theory. There are other approaches, one of them relying on C*-algebras, in which

Hilbert spaces appear as derived notions. The Feynman approach with path integrals is also very general and I will sometimes mention it. The Hilbert space approach has been preferred, however, because most textbooks use it and it is the best known. It also makes the statement of the principles particularly convenient, which is why I will use it in this book.

The Framework of the Theory

72. As a preliminary to the subtleties of quantum mechanics, many authors agree about what makes it in some sense extraordinarily simple: it is a linear theory. Two solutions of the Schrödinger equation, for instance, can be added to obtain another solution. Probability amplitudes combine linearly, so it is as if every problem in physics could be reduced to solving a set of linear algebraic or differential equations, although the simplicity is mitigated by the fact that the number of these equations is infinite. This linearity is important enough to be duly stressed, as Dirac did when he proposed it, under the name of the *superposition principle*, for it has precedence over every other.

The first principle I will state includes the superposition principle. It is more specific in so far as it determines the whole mathematical framework of the quantum theory. Many investigations on quantum mechanics have shown that it can be conveniently formulated in the framework of Hilbert spaces, except for some systems in statistical mechanics that formally include an infinite number of degrees of freedom. The superposition principle is reflected by the fact that a Hilbert space is a vector space and therefore linear.

Principle 1. *One can associate with every isolated physical system a definite Hilbert space. Every physical concept entering in the description or analysis of the system should be expressible in a mathematical language into which only notions involving this Hilbert space can enter.*

Complement: Hilbert Spaces*

I will assume the reader is aware of the elementary mathematical theory of Hilbert spaces, and the present complement is only meant for reference and notation, without any attempt at completeness. I will not give precise definitions nor rigorous statements and will remain at the conventional level usually found in physics textbooks.

A Hilbert space is a complex vector space with a scalar product. A vector space is a set whose elements are vectors and will be denoted by Greek letters. The vector sum $\alpha + \beta$ of two vectors α and β is also a vector, as well as the product $c\alpha$ of α by a complex number c. The dimension of space can be finite (when describing a spin for instance) or it can be infinite (in the case of complex-valued functions $\alpha(x)$, which are still called vectors).

A scalar product is an operation associating a (complex) number $(\alpha|\beta)$ to two vectors α and β. Scalar products are defined in mathematics by abstract rules, and two vectors are said to be orthogonal if their scalar product is zero. The norm of a vector α is the positive number $\|\alpha\|$ such that $\|\alpha\|^2 = (\alpha|\alpha)$. A vector is said to be normalized if its norm is 1. An orthonormal basis $(\alpha_1, \alpha_2, \ldots)$ is made of mutually orthogonal normalized vectors. It must be complete, so that any vector is a linear combination of the basis vectors.

One can also rely directly on the most important examples. For a Hilbert space with a finite dimension n, a vector is defined by its coordinates $(\alpha_1, \alpha_2, \ldots, \alpha_n)$ in a definite orthonormal basis and the scalar product is given by

$$(\alpha|\beta) = \alpha_1^*\beta_1 + \alpha_2^*\beta_2 + \cdots + \alpha_n^*\beta_n. \tag{8.1}$$

One recognizes the usual expression of a scalar product in a Euclidean space, except that the coordinates are complex. In the case of functions $\alpha(x)$ (a wave function, for instance), one writes

$$(\alpha|\beta) = \int \alpha^*(x)\beta(x)\, d^q x, \tag{8.2}$$

$$\|\alpha\|^2 = \int |\alpha(x)|^2\, d^q x, \tag{8.3}$$

q being the dimension of the space on which the variables x are defined.

An operator A is a linear mapping applying every vector α on another vector, which is denoted by $A\alpha$. This convenient definition does not cover however all the operators of interest in physics, and unbounded operators (representing for instance position or momentum) are only defined on a restricted domain. I will not be very cautious about these matters and I will rely on the idea that people aware of these difficulties will know how to avoid them or know the convenient references. Other people are either unaware or aware but careless or aware and expert enough to know that no serious problem can arise from these matters for interpretation.

In a given basis, an operator can be associated with a Heisenberg matrix with elements $(\alpha_j|A\alpha_k)$. Given an operator A, one can define another operator $A^\dagger$ such that $(\alpha|A\beta) = (A^\dagger\alpha|\beta)$, which is called its adjoint. A *Hermitian*, or *self-adjoint* operator is defined by $A = A^\dagger$. It is also called an *observable*.

Self-adjoint operators have a very important spectral property, which is the key to their importance in physics. Let A be one of them. There exists an orthonormal basis $\{\alpha_n\}$ consisting in eigenvectors of A such that

$$A\alpha_n = a_n \alpha_n, \tag{8.4}$$

the *real* number a_n being the eigenvalue of A associated with α_n.

This statement of the spectral theorem is correct for a finite-dimensional Hilbert space and I should only mention that when several eigenvalues coincide, one speaks of a degenerate eigenvalue. The case of an infinite-dimensional Hilbert space is much more involved and one may encounter discrete eigenvalues a_n or continuous ones. For instance, the operator representing the energy of a hydrogen atom has discrete eigenvalues E_n, representing the energy of a bound state. They are degenerate, and one must add two other quantum numbers l and m as indices for specifying a definite eigenvector (a stationary wave function). Every positive value E of the energy is also en eigenvalue, representing an ionized atom. The set of all the eigenvalues of a Hermitian operator is its spectrum. I will skip the usual discussion of normalization for the continuous spectrum with the use of delta "functions" and also the discussion of adjointness in the case of nonbounded operators.

Among the notions occurring in physics, are *traces*. In an orthonormal basis $\{\alpha_j\}$, the trace of an operator A is defined as the sum

$$\operatorname{Tr} A = \sum_j A_{jj}, \quad \text{with } A_{jk} = (\alpha_j | A\alpha_k). \tag{8.5}$$

This is well defined when the sum converges (strongly). The trace does not depend on the choice of a basis. Two properties are often useful, namely,

$$\operatorname{Tr} AB = \operatorname{Tr} BA, \tag{8.6}$$

$$\operatorname{Tr} ABC = \operatorname{Tr} CAB. \tag{8.7}$$

More generally, the trace of an operator product is invariant under a cyclic permutation of the factors.

Remarks

A few remarks can be made on the physical aspects of Principle 1. Physics is a pragmatic science, and the knowledge of a physical system is necessarily limited in its precision by the available experimental devices. This limitation has two main consequences concerning the isolation and the uniqueness of the system.

The principle mentions an isolated system, which means a system with no external influence and which is obviously an idealization. In practice it can be only a system that is momentarily isolated with good enough precision here and now, and not a system that has been isolated since eternity. It may have been prepared in a laboratory by means of some apparatus that is not acting any more. It can also be found in nature and observed. The system is not necessarily microscopic, and the principle can be applied to macroscopic objects, including the devices for preparation, conditioning, and measurements. To say that the system is really isolated is then, of course, an impossibility, and one can only assume that the interaction with the external surroundings is negligible or irrelevant.

The principle also mentions a definite (i.e., unique) system. Some people believe, on the contrary, that the probabilistic character of quantum physics is best expressed if one considers conceptually infinite ensembles of identical systems, as in statistical mechanics. This approach, however, must be excluded when one intends to include classical physics within the quantum framework. In that case, considering an infinite collection of copies of the solar system would certainly then be odd.

The Hilbert space to be associated with a system is a flexible setting rather than an absolute reference; for instance, when considering a hydrogen atom, one may use a Hilbert space for two particles—an electron and a proton. The setting is different when relativistic effects are taken into account and one uses the Dirac equation for the electron. Further improvements are found in quantum electrodynamics, when the Hilbert space includes every possible photon or electron-positron pair. One might also think of the quarks inside the proton, which would imply a still larger Hilbert space. In practice, one can always use the framework that is most convenient for the question one is investigating, up to a given level of accuracy.

The simplest way to characterize these various remarks about Principle 1 is to say that this principle defines the *mathematical language* of quantum theory.

RELATIVISTIC INVARIANCE

73. One needs a reference frame for describing a physical system. I will use only inertial frames with the assumptions of special relativity, although I will mainly restrict my considerations to nonrelativistic velocities. The relevant principle is

Principle 2. *The same laws of physics are valid in all the inertial reference frames. When applied to an isolated system, they can be expressed in the same Hilbert space.*

The definition of a reference frame does not belong to quantum mechanics. It involves the existence of a space-time with coordinates

$x^{\mu} = (\mu = 0, 1, 2, 3)$, which are ordinary numbers. A change of reference frame may involve a translation in the origins of space and time, a rotation of the space axes, and a Lorentz boost with an arbitrary velocity. It therefore depends on 10 parameters. The set of these transformations is called the Poincaré group (see, for instance, Weinberg, 1995). An element L of the group is a transformation of coordinates

$$x'^{\mu} = \Lambda^{\mu}_{\nu} x^{\nu} + \alpha^{\mu},$$

where the matrix Λ accounts for the rotation of space axes and the Lorentz boost, the four-vector a representing a translation of the origin of space-time. One does not consider the case when the orientation of the space axes or the direction of the time axis is changed.

It can be shown that Principle 2 implies an explicit form for the action of a reference change on the Hilbert space associated with a physical system. It is given by a unitary operator acting on vectors and operators, that is, on each vector

$$\alpha' = U(L)\alpha \tag{8.8}$$

the operators $U(L)$ being unitary and representing the Poincaré group so that $U(LL') = U(L)U(L')$.

*Note on the Proof**

The proof of this important result involves two steps, first a *Lemma*: the square of a scalar product $|(\alpha|\beta)|^2$ must be invariant in the transformation; and then a *Theorem* resulting in equation (8.8). The proof of the theorem is due to Wigner (1959), and it can be found in Weinberg's book (1995). The lemma is usually borrowed from the Copenhagen interpretation, and it must be reconsidered in the present approach. It can be derived from the representation of physical properties by projectors, which will be discussed in the next chapter.

*Mass and Spin**

The first chapters of Weinberg's book (1995) show how to construct the Hilbert space of a system consisting of several particles according to Principles 1 and 2. This means that the Hilbert spaces one has to deal with are well defined and their explicit construction does not need any further principle or assumption. It should be stressed particularly that nowhere does one refer to the historical discovery of quantum mechanics nor its process of quantization starting from classical physics. In the case of an isolated particle, one finds that relativistic invariance implies the existence of mass and spin, so that these notions and their mathematical description need no prior interpretation. These results were first obtained by Wigner and Bargmann (Wigner, 1939; Bargmann and Wigner, 1948). They have been extended to a nonrelativistic change of reference frame (Bargmann, 1954; Lévy-Leblond, 1963).

DYNAMICS

Observables

74. Before any statement of the law of dynamics, one must introduce the notion of a dynamical quantity, that is, an *observable*. The word "observable" refers to something that can in principle be observed and measured, and it thereby calls to mind an analogy with classical dynamical variables. By definition, however, an observable is, in the strict framework of Principle 1, an arbitrary self-adjoint operator. This strange relation between an abstract mathematical notion and a very intuitive physical concept raises an essential problem for interpretation. It is sometimes understated, for instance by von Neumann, who asserted as a matter of principle that one can always conceive of a device for measuring an arbitrary observable. I will not adopt this sweeping approach because measurements belong to interpretation and cannot intervene at the level of principles. The previous review of the history of interpretation in Part Two leads us to anticipate that a mastery of observables requires a fully developed interpretation. That is why at the present level of foundations I will consider an observable as a purely theoretical notion, namely, a self-adjoint operator.

Time Is a Variable and Not an Observable

Time, as it appears in Schrödinger's equation, is a space-time coordinate in some reference frame. It is therefore a pure number and not an observable. If one tried to associate it with an observable, one would arrive at an impossibility. It would be necessary to introduce a Hamiltonian for a perfect clock including time among its observables, the Hamiltonian being canonically conjugate to time. It can then be shown that a continuous time spectrum going from $-\infty$ to $+\infty$ implies an energy spectrum going down to $-\infty$. The clock is therefore necessarily unstable.

Dynamics

75. It will be convenient to express the fundamental law of dynamics as it was first formulated by Heisenberg:

Principle 3. *In a given reference frame, an isolated physical system is characterized by a Hamiltonian, or energy observable H. The time evolution of an observable A is given by the equation*

$$dA(t)/dt = (i/\hbar)[H, A(t)]. \tag{8.9}$$

Because observables are still considered presently as abstract quantities, the principle is necessarily formal. It introduces only a group of transformation acting on observables, with time as a parameter. Equation (8.9) is identical with equation (3.3) for the evolution of Heisenberg matrices. Schrödinger had replaced it with an equation for the wave function (or vector) ψ:

$$i\hbar\, d\psi/dt = H\psi. \tag{8.10}$$

Both versions are used in the usual practice of quantum mechanics. In Heisenberg's approach, the state vector is fixed and the observables evolve in time like classical variables $x(t), p(t)$. In Schrödinger's version, the state vector evolves and the observables are fixed.

*The Relation Between the Two Standpoints**

The mathematical equivalence between the two points of view is easy to show, *if one assumes the existence of a wave function.* Considering for simplicity the case when H does not depend on time, a formal solution of the Schrödinger equation (8.10) is given by

$$\psi(t) = U(t)\psi(0), \quad \text{with } U(t) = \exp(-iHt/\hbar). \tag{8.11}$$

But it is equivalent, in a Hilbert space, to let a unitary operator U act on the vectors while the operators are kept fixed or to keep the vectors fixed and let the operators change through a transformation $A \to U^{-1}AU$. This is exactly the relation between the two points of view from Schrödinger and Heisenberg. Taking into account the expression (8.11) for $U(t)$, the time evolution of

$$A(t) = U(t)^{-1}AU(t) \tag{8.12}$$

satisfies equation (8.9). Conversely, equation (8.12) can be considered as the solution of the differential equation (8.9) with the initial condition $A(0) = A$.

I chose Heisenberg's approach, which is slightly more complicated than Schrödinger's because it would be risky to pretend, before any interpretation, that one understands what a wave function is or even that it exists.

*Relations with the Classical Formalism**

There is a strong analogy between the quantum law (8.9) and classical Hamiltonian dynamics. In classical physics, the canonical variables $q(t)$ and $p(t)$ are time dependent and they obey the Hamilton equations (2.6). Poisson noticed that one can give a universal form to these equations for any dynamical variable $a(q,p)$. He defined the *Poisson bracket* of two functions $a(q,p)$ and $b(q,p)$ by

$$\{a,b\} = \sum_k \left(\frac{\partial a}{\partial p_k} \cdot \frac{\partial b}{\partial q_k} - \frac{\partial a}{\partial q_k} \cdot \frac{\partial b}{\partial p_k} \right). \tag{8.13}$$

The time evolution of a dynamical variable is then given by

$$\frac{\partial a}{\partial t} = \{h, a\}, \tag{8.14}$$

where $h(q,p)$ is the Hamilton function. The analogy between equations (8.9) and (8.13) is obvious.

One would still have to write out the Hamiltonian in order to make the physical foundations completely definite. Finding its exact form and the various possible interactions (gravitational, strong, weak, and electromagnetic) has been one of the main achievements of particle physics. When one is interested only in interpretation, one can be content with a nonrelativistic Hamiltonian for charged particles interacting via Coulomb potentials and with their coupling with photons. For an up-to-date expression of more general Hamiltonians, see for instance Weinberg, 1995.

*Hamiltonian and the Change of Reference Frames**

Using the Copenhagen measurement theory, one can show that the Hamiltonian occurring in equation (8.9) coincides with the operator generating time translations in the Poincaré group. I do not know a proof of this result in the present approach, but it does not really matter.

IDENTICAL PARTICLES

76. Matter and radiation are made of particles. Although one must use quantized fields for a complete description of particle physics, I will avoid their technicalities. One should, however, take into account the impossibility of distinguishing two identical particles (two electrons or two photons, for instance). One could very well introduce a principle stating this impossibility. Introducing an operator exchanging two identical particles as an observable, one would then refer its investigation to the interpretation of observables. I will not pursue this approach, and the relation between indistinguishability, symmetry, and spin will be stated globally as a unique principle.

The particles with an integral spin are called *bosons*, and those with a half-integral spin, *fermions*. Letting an index μ denote a value for the z-component of spin of a particle and p its momentum, one can assert

Principle 4. *The vectors in the Hilbert space of a system involving identical particles are completely symmetric under an interchange of the arguments* (μ, p) *for several identical bosons and completely antisymmetric for identical fermions.*

This principle restricts the Hilbert space to vectors that are either symmetric or antisymmetric under a permutation of particles. From a formal standpoint, the rule can be considered as a more precise definition of the Hilbert space. It implies, of course, a corresponding restriction on the observables, which must preserve the symmetry under a permutation.

The present principles of quantum mechanics can be considered, as complete and sufficient for obtaining an interpretation. Other basic notions, including the meaning of observables, the state of a system, and the role of probabilities, will be introduced later, after a minimal amount of interpretation.

ELEMENTARY LOGIC

77. Interpretation relies heavily on logic, whatever the kind of interpretation one prefers. I will therefore consider now the basic principles of logic. The formal (symbolic) character of quantum mechanics suggests a possible relation with some sort of formal logic, as von Neumann was first to notice. Some physicists have been reluctant to use formal logic, but computers have done much to change their prejudice. In fact, building an interpretation requires only the most elementary notions of logic.

Logic deals with *propositions* and the field of propositions to be considered in each application must be clearly specified. I will show later which fields enter into interpretation, but it is enough for the time being to assume that they exist. Three logical operations must be defined on a definite field of propositions. The first one is negation: every proposition is related to another one, which is its negation. There are also two operations, *and* and respectively *or*, combining two arbitrary propositions a and b to yield other propositions "a and b," or "a or b."

But the real power of logic lies in reasoning. To say "and," "or," or "no" is of little avail if one cannot say "therefore." This logical inference is a relation between two propositions, an antecedent a and a consequence b, which is written $a \Rightarrow b$ and expressed in words in various ways such as "a, therefore b"; "b, because a"; "if a, then b," and so on. There is also another relation, logical equivalence, which is written $a = b$ and is tantamount to the pair of inferences $(a \Rightarrow b, b \Rightarrow a)$.

These three operations and two relations are governed by a collection of axioms, which are the backbone of logic. For the sake of completeness, a brief list follows, but the reader can skip it without any inconvenience. They enter only into the sequel within some technical proofs that will not be given in full, anyway.

*The Axioms of Logic**

The axioms of equivalence are:

1. $a = a$;
2. If $a = b$ and $b = c$, $a = c$;

The axioms for negation (denoted by an upper bar) are:

3. $\bar{\bar{a}} = a$;
4. If $a = b$, $\bar{a} = \bar{b}$;

The axioms for *and* and *or*, where *and* is denoted by a point (.) and *or* by a plus sign (+):

5. $a.a = a$;
6. $a.b = b.a$; commutativity
7. $a.(b.c) = (a.b).c$; associativity
8. $a + a = a$;
9. $a + b = b + a$; commutativity
10. $a + (b + c) = (a + b) + c$; associativity
11. $a.(b + c) = a.b + a.c$ distributivity

The de Morgan rules relate *and* and *or* to negation:

12. $\overline{a.b} = \bar{a} + \bar{b}$;
13. $\overline{a+b} = \bar{a}.\bar{b}$;

The axioms for inference are:

14. The pair of implications $a \Rightarrow b$ and $b \Rightarrow a$ amounts to the logical equivalence $a = b$.
15. $a \Rightarrow (a+b)$;
16. $a.b \Rightarrow a$;
17. If $a \Rightarrow b$ and $a \Rightarrow c$, then $a \Rightarrow b.c$;
18. If $a \Rightarrow c$ and $b \Rightarrow c$, then $(a+b) \Rightarrow c$;
19. If $a \Rightarrow b$ and $b \Rightarrow c$, then $a \Rightarrow c$;
20. If $a \Rightarrow b$, then $\bar{b} \Rightarrow \bar{a}$.

In Boolean logic, the field of propositions is considered as a set S, and every proposition is associated with a subset of S. One can thus get a convenient graphical representation of logic in which the operations *and* and *or* can be replaced by the intersection and union of sets, and in which a negation $\bar{a}$ corresponds to the complement of the set associated with the proposition a. One can also introduce a universal proposition I associated with the set S and an empty proposition 0 associated with the empty set (the notation anticipating the fact that these two propositions will be represented later by the identity operator and by zero).

THE AXIOMS FOR PROBABILITIES

78. The axioms that probabilities must satisfy are very important in spite of their simplicity. There are only three axioms for the probabilities $p(a)$, which are defined on a set of events or a field of propositions (notice that in ordinary probability calculus, every event can be described by a sentence and is therefore expressed by a proposition).

Axiom 1 (positivity).

$$p(a) \geq 0. \tag{8.15}$$

Axiom 2 (normalization).

$$p(I) = 1, \tag{8.16}$$

where I is the universal proposition allowing for every possibility.

Axiom 3 (additivity). *If two propositions a and b are mutually exclusive (i.e., "a and b" = 0, representing the empty proposition or an inexistent event), then*

$$p(a \text{ or } b) = p(a) + p(b). \tag{8.17}$$

One may recall finally that the *conditional probability* for a proposition b if given another proposition a (with $p(a) \neq 0$) is defined by

$$p(b|a) = p(a.b)/p(a). \tag{8.18}$$

9

Quantum Properties

Long ago von Neumann had a remarkable idea, which was the germ of a universal language for the interpretation of quantum mechanics. He associated with a quantum property of a system (or an elementary predicate, as he called it) a projection operator in Hilbert space, or a *projector,* as will be said here for brevity. The elements of this language will now be given.

THE COMPONENTS OF AN EXPERIMENT

79. When one describes a quantum experiment, it may be convenient for the purpose of interpretation to distinguish various parts in the experimental apparatus. One can consider three of them, which are used for preparing, conditioning, and measuring atoms, particles, or microscopic objects.

In ordinary interference experiments, for instance, the preparation device is a source of light, often a laser. The conditioning device is an interferometer (which can be simply a screen with two slits). The measuring device can be a photographic plate, a set of photodetectors, or simply the retina of an observer. In nuclear and particle physics, the preparing device is most often an accelerator. Conditioning can involve beam guides, targets, and electronic devices. Measurements are made with detectors.

This distinction is just a convenience for discussion and it is not crucial if some device is considered as belonging to one of the constituent parts or another. It does not matter whether a polarizer in front of an interference experiment is said to belong to the preparing or to the conditioning parts of the apparatus, although distinguishing them in our language is sometimes useful.

THE NOTION OF PROPERTY

80. Physics already has its own plain language, which is used by every physicist in lectures, articles, and meetings. For instance, one often describes an interference experiment in the following manner: A pho-

ton enters the interferometer. A moment later its wave function is divided into two parts, which are localized in the two arms of the interferometer. These two parts of the wave function then add up together behind the interferometer. The photon is finally detected at a specific place by a photodetector or an emulsion grain.

Similar sentences enter into the description of an experiment, such as, "A reaction $n + p \rightarrow d + \gamma$ has occurred"; "A proton leaves the accelerator with a velocity v, within an error bound Δv"; "When exiting the Stern-Gerlach device on this trajectory, the atom has a spin z-component equal to $+1$". Although these various assertions look rather different, all of them refer to something occurring at a definite time in a definite experiment. I will say that they describe *properties* of a microscopic object (an atom or a particle), although I will refrain from defining more precisely right away what is a property in general. For the time being, I wish only to make clear that the properties we are considering are part of a language. They can express various levels of reality or virtuality: something which occurs, which might occur, or simply something we want to talk about. The name "elementary predicate," which was used by von Neumann, is perhaps clearer than "property" from this standpoint, but it may look a bit pedantic. In any case, it should be clear that the properties belong to the language of physics with no pretence of always following reality exactly.

Quantitative Properties

81. Many properties have a similar pattern. They involve an observable A, a real set Δ (belonging to the spectrum of A), and they state a property, such as, "The value of A is in Δ." The observable A can be a position, a momentum, an energy, or almost anything. The set Δ can be an interval (for a velocity component for instance) or a single value. To say, for instance, that an atom is in its ground state is to state a property, in which A is the energy observable and Δ is the singlet set $\{E_o\}$ consisting exactly of the ground state energy E_o. When several observables commute, such as the three coordinates of a particle position, they can be joined into a multidimensional observable while Δ becomes a multidimensional domain in their combined spectrum. For instance, when saying that a particle is in a space domain V, one can introduce a three-dimensional position observable and then state: "the value of X is in V."

All the properties of that type may be called *quantitative properties* insofar as they assert the value of an observable.

82. Von Neumann's key idea was to associate a quantitative property of a system with a projection operator in the corresponding Hilbert

space. It is well known that a proposition can be coded, especially mathematically. This is exactly what occurs when a message, a musical piece, or a record is registered numerically as a series of bits 0 and 1. In the present case, the properties of relevance in physics are registered as a projection operator, and this expression is directly related to the basic quantum principles. It fits perfectly the first principle of the mathematical framework of the theory, which was discussed in Chapter 8.

The idea of a projection operation first occurred in ordinary Euclidean geometry. Considering, for instance, a plane M in three-dimensional space containing the origin, every vector v can be written in a unique way as $v = v_1 + v_2$, where v_1 belongs to M and v_2 is orthogonal to M. One says that v_1 is the projection of v on M. One goes from v to v_1 by a linear operation, that is, a projection operator E such that $v_1 = Ev$. A vector in M is its own projection so that one has $Ev_1 = v_1$ and consequently $E^2 = E$.

This can be extended from a three-dimensional Euclidean space to a complex Hilbert space with finite or infinite dimension and an arbitrary subspace M (a mathematician would only insist on the fact that M must be a "closed" space). The corresponding projector E satisfies the relation $E^2 = E$, and it is self-adjoint.

A projector is defined generally as a self-adjoint operator satisfying the relation

$$E^2 = E. \tag{9.1}$$

Being self-adjoint, E has a real spectrum. The condition (9.1) shows that every eigenvalue e satisfies the equation $e^2 = e$, so that it can only be 0 or 1. The eigenvectors with eigenvalue 1 span a subspace M of the Hilbert space, and those with eigenvalue 0 are orthogonal to M. Conversely, to any (closed) subspace M corresponds a projector E.

Let us then consider the quantitative property "the value of A is in Δ." One can introduce a basis of eigenvectors of A, say $|an\rangle$, where a denotes an eigenvalue, either discrete or continuous, and n is a degeneracy index. Denoting with Dirac by $|a\rangle\langle a|$ the projector on the one-dimensional Hilbert space generated by a vector α, one can introduce a projector E which is directly associated with the property, by

$$E = \sum_{a \in \Delta} \left(\sum_n |an\rangle\langle an| \right) \tag{9.2}$$

One can take Δ to be a subset in the spectrum of A as was done previously, or one can consider directly a real set. If in that case there is no eigenvalue of A in Δ, one has $E = 0$. If the whole spectrum of A is included in Δ, then $E = I$. The sum in equation (9.2) is an integral in the case of a continuous spectrum, with the usual normalization conventions. I will not discuss the mathematical subtleties associated with the existence of this integral, which implies essentially that the intersection of Δ with the spectrum must be a so-called Borel set.

Example

One considers the position X of a particle as an observable, the domain Δ being a three-dimensional volume V. The corresponding projector is

$$E = \int_V |x\rangle\langle x|\, dx.$$

When acting on a Hilbert space vector α associated with a wave function $\alpha(x) = \langle x|\alpha\rangle$, it gives a vector $\beta = E\alpha$ such that

$$\beta(y) = \langle y|\beta\rangle = \int_V \langle y|x\rangle\langle x|\alpha\rangle\, dx.$$

Using $\langle y|x\rangle = \delta(x-y)$, one finds that $\beta(y)$ is identical with $\alpha(y)$ when y belongs to V and it vanishes outside V.

With every quantitative property, one can thus associate a projector. Conversely, every projector E corresponds to at least one property as one sees by taking E itself as an observable and the singlet set $\{1\}$ for Δ.

SEMANTICS

83. Semantics is the science of meaning. When considering properties, one may notice that many of them can be differently worded but nevertheless associated with the same projector E. All of them, however, are logically equivalent so that a language using properties has a perfectly well-defined meaning. Projectors can therefore afford a more faithful representation for the meaning of a property than any other language, ordinary words, particularly can do.

Consider as an example two simple properties. According to the first one, "the value of A is in the interval $[-1, +1]$," whereas the second one states "the value of A^2 is in the interval $[0, 1]$." They obviously have the same meaning if they have any. As a matter of fact, they are associated with the same projector, since $|a, n\rangle$ is both an eigenvector of A with the eigenvalue a, and an eigenvector of A^2 with the eigenvalue a^2.

*The General Case**

Let f be a function defined on the spectrum of A (it is supposed to be Lebesgue integrable, but this is true of any honest function). It maps the subset Δ of the spectrum of A on the real set $\Delta' = f(\Delta)$. Let one assume that no point of the spectrum outside of Δ is mapped in Δ'. The two properties stating respectively "the value of A is in Δ" and "the value of $f(A)$ is in $f(\Delta)$" can then be shown to have the same projector. There are therefore infinitely many properties that are equivalent to a given one.

Finally, a property will be defined in general by its mathematical expression, so that a property is any statement that can be expressed (coded) by a projector.

Taking Time into Account

84. Most properties are supposed to hold at a precise time during some step of an experiment. They are therefore more properly expressed by a statement such as "the value of $A(t)$ is in Δ," involving the time-dependent observable $A(t) = U^{-1}(t)AU(t)$. The corresponding projector will be denoted by $E(t)$. It can be easily related with the projector E for the timeless property "the value of A is in Δ." If one notices that $U^{-1}(t)|a, n\rangle$ is an eigenvector of $A(t)$ with the eigenvalue a, the definition (9.2) gives

$$E(t) = U^{-1}(t)EU(t). \tag{9.3}$$

OTHER EXPRESSIONS FOR PROPERTIES

85. Many statements are used in the language of physics even though they do not look quantitative at first sight. We already noticed that an apparently nonquantitative property such as "the atom is in its ground state" can be replaced by a quantitative property relative to the atom energy. There are other examples. Suppose one wants to say that a reaction $n + p \to d + \gamma$ has occurred. One would say more exactly that at some time t, the system is $n + p$, whereas at a later time t' it is $d + \gamma$. Both statements are in fact properties because they can be expressed by definite projectors. In order to describe the reaction, one must consider two different Hilbert spaces, $\mathcal{H}_{\mathrm{np}}$ for the two nucleons and $\mathcal{H}_{\mathrm{d}\gamma}$ for the deuteron and photon. The full Hilbert space is a direct sum $\mathcal{H} = \mathcal{H}_{\mathrm{np}} \oplus \mathcal{H}_{\mathrm{d}\gamma}$, the subspace $\mathcal{H}_{\mathrm{np}}$ is associated with a projector E_{np} express-

ing the first property, and a projector $E_{d\gamma}$ on $\mathcal{H}_{d\gamma}$ represents the second property. The method works for any reaction, including chemical ones.

Another interesting example occurs in the case of an interference experiment. One may assert "the particle wave function is ψ." There is again a projector which is $|\psi\rangle\langle\psi|$, expressing that statement. This does not mean that the particle is described "really" by this wave function. It is only a proposition one may wish to express for the sake of an argument. It belongs to the language of interpretation.

The conclusion is that any statement one may introduce when describing a microscopic system, including atoms and particles or radiation, can be expressed by a projection operator. This assertion cannot be proved, but it is a matter of experience, because no exception has ever been found.

This "Neumannian" language has two essential virtues, which will appear later. It is universal, which means that it does not cover only the properties of microscopic quantum objects, but as we shall see, it can also be extended to the classical properties of a whole experimental device. It is also sound, that is, it is provided with a definite logical framework. Some work will be needed for establishing these two basic qualities, and it will be my task in the next few chapters to do so.

One should finally mention von Neumann's analogy between projectors and the "truth value" of a property in logic. A logical proposition is often considered to have a truth value, which means that it can be true or false. Similarly, an eigenvalue of a projector can only be 1 or 0. The analogy is tempting but one must be very careful. The notion of truth can be used universally in mathematics, but it is rather different in a natural science, where it is primarily associated with the reality of a fact. Since we are still far from anything analogous to facts or truth, I will make no use of von Neumann's remark.

ABOUT OBSERVABLES

86. The present section lies somewhat outside the main topic, but it may be of interest in shedding some light on the notion of observable. As a matter of fact, an observable lies somewhere between a classical random variable and an ordinary classical dynamical variable, which entails looking at the observable from two different standpoints.

Let us first keep to the domain of language, or of ideas as opposed to practice and reality. An ordinary classical variable is described by a number, which is supposed always to have a well-defined value x. A random variable is less simple. Its definition involves three distinct notions: (1) There are some propositions e_j, which express possibilities rather than actual events. As a possibility, each one of them belongs to language (I suppose their number to be finite and labelled by an index j for simplicity). It may state, for instance, that a lottery wheel can stop in some orientation in a definite sector. (2) A numerical value x_j of a classical dynamical variable is associated with each possibility (i.e., each proposition). It may be an angle or a number written on a sector of the wheel, but I will use the angle. (3) A probability p_j is assigned to each proposition (possibility).

Some of these characteristics also appear in the case of a quantum observable. There are also possible numerical values x_j (the eigenvalues) and also propositions e_j, which are presently expressed by projectors E_j. The probabilities do not belong properly however to the observable. They depend on the state of the system and are nearer the reality of physics, i.e., for instance, how the wheel is started and how it rolls. An observable is therefore basically a collection (x_j, E_j), and this is clearly a part of the language of properties. It could have been used as a definition from which this language would have been constructed.

What is extremely remarkable is that a collection (x_j, E_j) can be used to construct an operator $X = \Sigma_j x_j E_j$, which can be in correspondence with a standard classical dynamical variable. This dual feature of observables suggests a remark about their basic nature and recalls Bohr's distinction between the mathematical symbols of quantum theory and classical concepts (which are supposed much richer in their associative power). This is indeed a very deep idea, but as Bohr also said once (perhaps as sort of a joke), the contrary of a deep truth is another deep truth. Quantum observables possess an extraordinary associative power in their conjunction of two rich classical concepts. They are by themselves a concept, perhaps the essential one in physics. The difficulty in understanding them is due to their wealth, which exceeds common sense. As long as they have not become a part of our brain, quantum physics cannot be really understood.

SUMMARY

Every description of the microscopic events occurring in an experiment can be expressed by quantum properties, which are expressed mathematically by projection operators (projectors) in Hilbert space. A property may state that the value of an observable is in some real range of numbers at a given time, but many other useful statements can also be expressed in the same language.

10

Classical Properties

The language of properties cannot pretend to be universal if it disregards the most obvious components of an experiment, that is, everything macroscopic in it. Von Neumann attempted the necessary extension, but with no convincing success. The program was only realized much later when new theoretical methods had been found and new advances in mathematics provided the necessary tools. I will explain them in the present chapter and show how one can speak of classical physics while relying only on the basic principles of quantum mechanics.

CLASSICAL PROPERTIES

87. Classical physics is used in the description of a quantum experiment in at least two ways. There is the description of the apparatus, which need only refer to the collective coordinates—what can be seen and directly controlled. It is also often convenient to speak of an atom or a particle as though it had a classical trajectory, as though its quantum features did not matter. This is how one thinks of the electrons in a TV tube or of the particles in an accelerator, and as a matter of fact, the conception of an experiment relies heavily on this kind of commonplace considerations.

The idea of speaking jointly of the position and momentum of a particle is clearly at variance with the uncertainty relations. But Planck's constant is so small that it is most often negligible when compared with the uncertainties inherent in the working of an apparatus. This means that one does not need absolute classical properties, which would state that the position and momentum of a particle have exactly the values (x_o, p_o) with an infinite number of digits. It is enough to state that the particle position and momentum have the values (x_o, p_o) up to upper error bounds $(\Delta x, \Delta p)$. There should certainly be some way of justifying such a reasonable statement when $\Delta x . \Delta p \gg \hbar$.

It will be convenient to use a graphical representation for such a classical property (see Figure 10.1). Considering for simplicity a unique coordinate x, one can associate the property with a rectangle R in

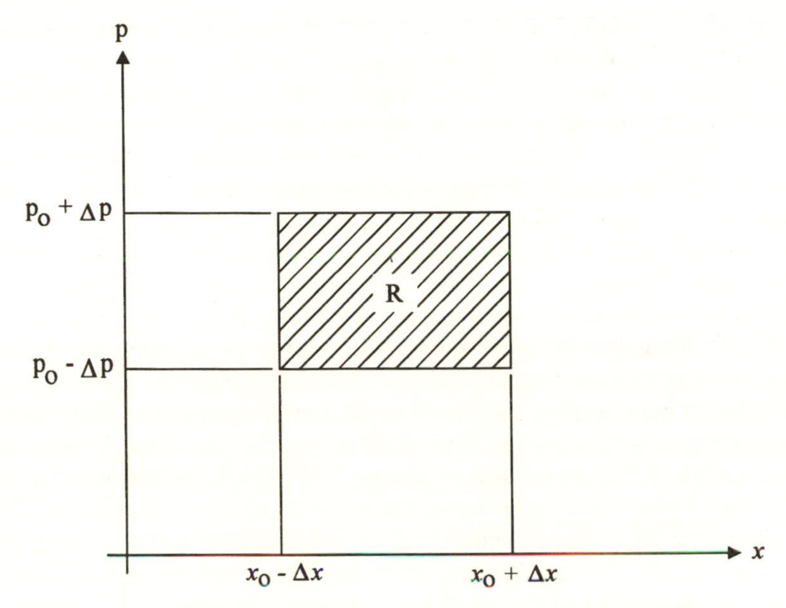

Figure 10.1. Representation of a classical property. A rectangle R in phase space with the sides $2\,\Delta x$ and $2\,\Delta p$ represents a classical property asserting simultaneously position and momentum (x_o, p_o) within given error bounds $(\Delta x, \Delta p)$.

phase space, whose center has coordinates (x_o, p_o) and the sides are $(2\,\Delta x, 2\,\Delta p)$. The key question one must answer is how to put this classical property on the same footing as a quantum one, i.e., to associate it with a projector in the Hilbert space of the particle. This question is essentially a purely mathematical one: can one associate a projection operator in Hilbert space with the rectangle R?

CLASSICAL DYNAMICAL VARIABLES

88. The quantities (x, p) that have been introduced are so familiar that they look natural. Old habits, however, are not a good argument when one wants to put interpretation on a firm ground. Once the principles of quantum mechanics have been stated, every new notion must rely on these principles and on them only. This is not trivial and the approach to classical physics that was used by Ehrenfest in 1927 is an example of the difficulties. He considered (x, p) to be the average values of the observables (X, P) with respect to a convenient wave function. The trouble is that the method cannot be extended to a general observable

A because its average value a is not a function $a\,(x, p)$, as expected for a genuine dynamical variable in classical mechanics.

The crux of the matter is to find a way for directly associating a function $a(x, p)$ with a more or less arbitrary quantum observable A. Intuitive physical considerations may help us in finding this correspondence between operators and functions (the word "correspondence" being not, of course, chosen ingeniously). The function associated with the position observable X should reduce to x, and the momentum observable P to p.

The correspondence must also be linear, that is, it must conserve the sum of physical quantities and their product by a number. This is because many physical quantities can be added together, such as energies, momenta, or angular momenta. A multiplication by a number is equivalent to a change of units (using feet in place of meters, for instance), and one expects these operations to have the same effect on quantum and classical quantities.

This is not enough however for fixing completely the correspondence. One needs other conditions, of which many variants are possible. One may ask for an exact correspondence between the quantum and classical motions of a harmonic oscillator, or one may demand a correspondence of the classical canonical change of coordinates and the quantum unitary transformation occurring under such a motion. One may also go back to Dirac's method and ask for the best possible correspondence between quantum commutators and Poisson brackets.

Most reasonable criteria of that kind lead to a specific correspondence, which was discovered for a special case by Wigner (1932) and made systematic by Hermann Weyl (1950). It associates the observable X with x and more generally a function $f(X)$ with $f(x)$ and a function $g(P)$ with $g(p)$. It can be extended to non-Hermitian operators, which are associated with complex functions of (x, p). The correspondence formula is given explicitly by

$$a(x, p) = \int \langle x + y/2 | A | x - y/2 \rangle \exp(-ip.y/\hbar)\, dy, \qquad (10.1)$$

whatever the dimension n of the configuration space (i.e., the spectrum of the X's).

Taking into account the expression of plane waves,

$$\langle x | p \rangle = (2\pi\hbar)^{-n/2} \exp(ip.x/\hbar), \qquad (10.2)$$

one can check easily the correspondence between the observables $(I, X, P, XP_y - YP_x, P^2/2m + V(X))$ and the functions $(1, x, p, xp_y - yp_x, p^2/2m + V(x))$ for the most familiar quantum quantities and their classical expressions.

One may notice that equation (10.1) is a Fourier transform replacing the variable y by p. The inverse Fourier transform provides an expression for the matrix elements of an operator A in terms of the classical function $a(x, p)$:

$$\langle x'|A|x''\rangle = \int a((x'+x'')/2, p)\exp(-ip.(x''-x')/\hbar)\, dp(2\pi\hbar)^{-n}. \tag{10.3}$$

The (nonnecessarily real) function $a(x, p)$, which is associated with the (nonnecessarily self-adjoint) operator A by equation (10.1) is called the (Weyl) *symbol* of the operator. Also, for reference, the action of an operator A on a function $\psi(x)$, that is, $\phi = A\psi_j$, is given in terms of its symbol $a(x, p)$ by

$$\phi(x) = \int a\left(\frac{x+y}{2}, p\right)\exp[ip.(x-y)/\hbar]\psi(y)\, dy\, dp(2\pi\hbar)^{-n}. \tag{10.4}$$

The discovery of these "magic" formulas inaugurated a return of quantum mechanics to its origins. The correspondence between classical and quantum physics no longer needs to be guessed from some quantization process; on the contrary, classical physics is considered to emerge from a more fundamental quantum physics. To obtain classical physics, one must in some sense "dequantize" or "classicize" quantum mechanics.

Mathematicians discovered for their own purpose many important consequences of these formulas within the modern theory of linear operators. This very efficient theory, which was developed between the 1960s and the 1980s and in which research is still quite active, is called *microlocal analysis* or *pseudo-differential calculus*. Its powerful techniques are most efficient when one has to justify rigorously some of the foundations of interpretation. They are not easy however, and I prefer more heuristic methods for illustration.

*Note 1.** One should mention a remarkable versatility of the microlocal methods: the use of Fourier transforms in the basic formulas could lead to the conclusion that the configuration x-space must be a Cartesian space R^n. But the word "microlocal" in *microlocal analysis* means that this technique can be adapted to practically any configuration space, whether it be infinite or finite, plane or curved, so that the corresponding operations of analysis are "localized" in phase space.

*Note 2.**

For mathematically inclined readers, the "function" $a(x, p)$ associated with a bounded operator is, in general, a distribution. A symbol is more exactly an infinitely differentiable function of x and p with some explicit bounds for the derivatives. The corresponding operator is said to be pseudo-differential. This class includes the differential operators of mathematical physics and their Green functions as well as many other interesting operators.

CLASSICAL PROPERTIES AND PROJECTION OPERATORS

89. I want to return to the question I asked at the beginning: how can one associate a projector E_R with a rectangle R in phase space in order to give a quantum meaning to a classical property? Two different methods have been used for that purpose: either microlocal analysis or the so-called *coherent states* in an approach first proposed by Hepp (1974a). I will begin with the second method, whose notions are more easily grasped with a physics background, whereas the first method, more powerful, is more abstract.

"Coherent state" is only a conventional name for a normalized Gaussian wave function. Let such a function $g_{qp}(x)$ correspond to average values $\langle X \rangle = q$, $\langle P \rangle = p$ for (one-dimensional) position and momentum. Let σ be the uncertainty for X and σ' for P (with $\sigma . \sigma' = \hbar/2$, since this is a Gaussian wave function). More explicitly,

$$g_{qp}(x) = (2\pi\sigma^2)^{-1/4} \exp\left[-(x-q)^2/(4\sigma^2) + ipx/\hbar\right].$$

When von Neumann tried to answer the question we are considering, he started from the idea that a definite number N of quantum states is associated with the rectangle R. One may think of R as a box with reflecting or periodic boundary conditions in x in which a "particle" is moving with specific bounds on p. N is the number of semiclassical states in the box, which is familiar in elementary statistical mechanics. Von Neumann chose N Gaussian wave functions for representing these N states with their centers (q, p) located at the vertices of a regular lattice. His results were not too good, however, and Hepp obtained much better ones by smoothing out the lattice and using a continuous distribution of Gaussian functions. He introduced an operator

$$F = \int_R |g_{qp}\rangle\langle g_{qp}| \, dq \, dp (2\pi\hbar)^{-1}, \tag{10.5}$$

which will be the candidate here for the role of a projector.

This operator has many encouraging features. It is Hermitian, with only discrete eigenvalues f_j. All the eigenvalues are located in the interval [0, 1]. Some of them are very close to 1, and their number is essentially N, which is the area of the rectangle with Planck's original constant h as a unit:

$$N = 4\,\Delta x\,\Delta p/h = \text{Area}(R)/(2\pi\hbar) = \int_R dq\,dp(2\pi\hbar)^{-1}. \quad (10.6)$$

This relation suggests that the number of quantum states associated with R is equal to N, a result that is well-known and originated with Planck himself. The remaining eigenvalues of F can be split in two parts. Most of them are extremely close to 0 and, finally, some of them, whose number is approximately $\sqrt{N}$, are dispersed between 0 and 1.

Were it not for these relatively few intermediate eigenvalues, F would be essentially a projector in Hilbert space, the subspace on which it projects having dimension N. The fact that the operator F is physically associated with the rectangle R is shown by its action on wave functions. If one considers a wave function with average values of X and P well inside (or respectively outside) R and uncertainties $(\Delta x, \Delta p)$ small enough, it is practically an eigenfunction of F with the eigenvalue 1 (respectively 0). Altogether, these results seem to imply that we are not too far from our goal.

*Proof.**

The proof of these results is not too difficult and I will therefore sketch it, although I will not discuss all the details because some explicit calculations must be made. They always are either an explicit computation or an estimate of a Gaussian integral, both of which belong to the standard tools of mathematical physics. They can become rather long when discussed in detail, so I will therefore state only the result of these calculations. For instance, one can obtain directly the scalar product of two Gaussian wave functions

$$\langle g_{qp}|g_{q'p'}\rangle = \exp[-(q-q')^2/2\sigma^2 - (p-p')^2/2\sigma'^2]. \quad (10.7)$$

It is convenient to choose σ and σ' respectively proportional to the sides of the rectangle R so that $\sigma = \varepsilon\,\Delta x$ and $\sigma' = \varepsilon\,\Delta p$. Since $\sigma.\sigma' = \hbar/2$, one has $\varepsilon = (\hbar/2\,\Delta x\,\Delta p)^{1/2}$. The important parameter ε will be found to control all the approximations.

The symbol $f(x,k)$ of F can be computed, according to equation (10.1), in terms of integrals and Fourier transforms of Gaussian functions, from which one obtains

$$f(x,k)=\int_R \exp[-(x-q)^2/2\sigma^2-(p-k)^2/2\sigma'^2]\,dq\,dp(2\pi\hbar)^{-1}. \quad (10.8)$$

It is clear that F is self-adjoint. It is also positive definite because one has for an arbitrary wave function ψ,

$$\langle\psi|F\psi\rangle=\int_R |\langle g_{qp}|\psi\rangle|^2\,dq\,dp(2\pi\hbar)^{-1}\geq 0. \quad (10.9)$$

In order to show that the spectrum of F is discrete and bounded by 0 and 1, one needs the traces of F and F^2. One finds easily that

$$\mathrm{Tr}\,F=\int_R dq\,dp(2\pi\hbar)^{-1}=N. \quad (10.10)$$

Using the scalar product (10.7), one can show that $\mathrm{Tr}\,F^2$ is a bounded quantity (very close to N). But self-adjoint operators whose square has a finite trace are known as Hilbert-Schmidt operators, and their spectrum is discrete.

According to equation (10.9), the eigenvalues of F are nonnegative. When the integral (10.5) is extended to the entire phase space rather than being restricted to R, one obtains the identity operator (the simplest way to show this is to notice that the symbol (10.8) becomes identically equal to 1, which is the symbol of identity). The operator $I-F$ is therefore given by an expression similar to (10.9) integrated outside R. Using a previous argument, this shows that the eigenvalues of $I-F$ are positive. Denoting by f_j the eigenvalues of F, one has therefore $0\leq f_j\leq 1$.

One can investigate the distribution of the eigenvalues in more detail. It was already noted that $\mathrm{Tr}\,F=N$ and $\mathrm{Tr}\,F^2$ is close to N. The relation $\mathrm{Tr}(F-F^2)=\Sigma_j f_j(1-f_j)$ shows that the quantity $\mathrm{Tr}(F-F^2)$ is a good measure for checking how much F differs from a projector, since it would be zero if all the eigenvalues were strictly equal to 1 or 0. A more precise estimate shows that

$$\mathrm{Tr}(F-F^2)=NO(\varepsilon), \quad (10.11)$$

with the previous value of ε. One can deduce from these estimates the properties mentioned for the distribution of eigenvalues.

This distribution can be confirmed by using another example in which the rectangle R is replaced by an ellipse whose boundary is given by

$$\frac{x^2}{\Delta x^2}+\frac{p^2}{\Delta p^2}=1.$$

All the previous results are unchanged, except that the value of N is now related to the ellipse area as in equation (10.6). One can, moreover, compute explicitly the eigenvalues. Introducing an oscillator Hamiltonian,

$$H = \frac{X^2}{\Delta x^2} + \frac{P^2}{\Delta p^2},$$

it can be shown that H and F commute (in view of the simple action of H on a Gaussian wave function g_{qp}). The eigenfunctions of the two operators must therefore coincide, and they are explicitly known for H from the theory of harmonic oscillators. Let them be ψ_n. The eigenvalues of F can be obtained by an explicit computation of $\langle \psi_n | F\psi_n \rangle$, from which one obtains

$$f_j = (1/j!)\int_0^N t^j \exp(-t)\, dt.$$

This integral (the incomplete gamma function) is also well known. It can be studied analytically or numerically. A look at its expression shows immediately that about N eigenvalues are close to 1, about $\sqrt{N}$ of them are located between 0 and 1, and the rest of them are very close to 0. Finally, the properties I mentioned for the eigenfunctions of F are easily estimated, at least for Gaussian wave functions whose center is well inside (or outside) the rectangle or the ellipse.

90. The result obtained is so close to our goal that it is worth looking at its shortcomings in order to use it better. We have not really obtained, for instance, a projector with all its eigenvalues strictly equal to 1 or 0, but a relatively small fraction of the eigenvalues of F lie between 0 and 1 without being close to 1 or 0. We shall say that F is a *quasi-projector*, a well-known notion in mathematics.

One can easily remedy this shortcoming for obtaining a true projector in the following manner: one does not change the eigenfunctions of F, and one changes the eigenvalues f_j. Those belonging to the interval $[1/2, 1]$ are replaced by 1 and those in the interval $[0, 1/2]$ by 0. I will call this operation a straightening of F. It does not appreciably modify the properties of the wave functions microlocalized inside (or outside) R, and one can thus obtain a true projector associated with R.

This procedure is not above criticism, and there is much arbitrariness in the process of straightening. Why sort out the eigenvalues according to their being larger or smaller than 1/2. Why not put the cutoff at 1/3? Why not change slightly the eigenvectors? Since the starting point of equation (10.5) was no more than a lucky guess, one gets the impression that the answer cannot be unique. There must be a whole family of projectors rather than a unique one, that is associated with R.

Simple physical arguments also support the idea of a multiple answer. To begin with, it is clear that one would not seriously consider associating a projector with a point in phase space because of the uncertainty relations. But, after all, the boundary of R was sharply specified, and this is also in conflict with these relations. This can be seen if one deforms the boundary slightly by giving slight undulations to the sides of the rectangle, indenting them, or smoothing out the corners. These changes cannot be significant from a quantum standpoint as long as they are small when compared with Planck's constant. Wave mechanics is too shortsighted in its vision of phase space for seeing such minute details. This lack of precision should be reflected by a fuzziness in the definition of classically meaningful projectors or, equivalently, by their multiplicity and their approximate character. This remark sheds a new light on our basic question, which should now consist in identifying a whole set of projectors as the right answer. One must also find criteria for expressing the equivalence of different projectors in the set if they are all on the same footing for expressing the same classical property.

91. This is where the full power of microlocal analysis enters (Omnès, 1997a), and one proceeds as follows: One begins by constructing a set of quasi-projectors by starting from their symbols. Let $f(x,p)$ be the symbol of some quasi-projector F. One avoids a boundary that is too precise by choosing a smooth function $f(x,p)$ that goes gradually from the value 0 outside R to 1 inside. This transition may occur in a neighborhood of the boundary, which will be called the *margin* of R (see Figure 10.2).

With estimates borrowed from mathematics, it can be shown that the best results (i.e., the smallest errors, which will be given shortly) are obtained when (1) The function $f(x,p)$ is infinitely differentiable and obeys some peculiar bounds on its derivatives on which I will not insist, and (2) the previous parameters σ and σ' are taken (in order of magnitude) for the widths of the margin along the p and x-sides respectively. Perhaps this is already too technical to be really interesting and it can be disregarded, provided one remembers that the *set* of quasi-projectors is essentially well defined, the previous operator (10.5) being one of them.

The properties of the new quasi-projectors are very similar to those of our first example. The spectrum is discrete. The eigenvalues lie essentially in the interval $[0,1]$. They accumulate close to 0 or 1 as before. The exact width of the margin enters into the precise spectral properties, and it cannot be completely arbitrary. If it is too narrow, some eigenvalues lie outside the interval $[0,1]$ and the F operator does not look much like a projector. If the margin is too wide, there are too many eigenvalues dispersed between 0 and 1, and one loses a lot of precision in the definition of the classical property one is considering. This is the reason for the optimum width.

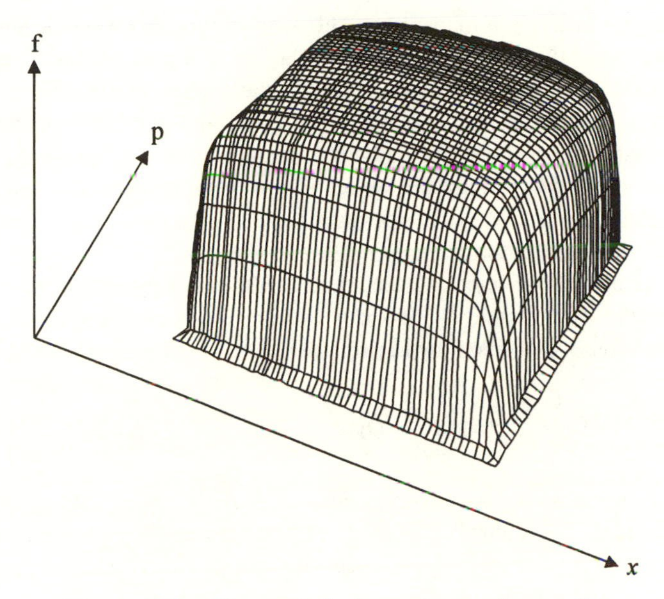

Figure 10.2. The symbol of a projector. A smooth function $f(x, p)$ is equal to 1 or 0 except in a margin of the classical rectangle R where it goes gradually from 0 to 1. It is the symbol of a quasi-projector representing the corresponding classical property, from which one can obtain a true projection operator by straightening the eigenvalues.

There are as many quasi-projectors in the set as functions $f(x, p)$ satisfying these rough conditions. All of them have essentially the same trace, namely N. All of them are close together, as expected if their meaning is equivalent.

*The Topology of Projectors**

Some topology must enter into the discussion of equivalence since it is required for any consideration of vicinity. I will indicate what it looks like for the readers who wish this kind of information although it is not essential for understanding the basic ideas. One wants to characterize the distance between two quasi-projectors F and F'. The convenient topology is the *trace norm* topology. It relies on the "absolute value" of $F - F'$, which is the positive operator $|F - F'|$ whose square is equal to $(F - F')^2$. The trace norm of $F - F'$ is defined as the trace $\mathrm{Tr}|F - F'|$. It can be shown that it is a distance between operators (i.e., it is positive and satisfies the triangle inequality). This kind of topology was not much used in operator theory because the calculation of a trace norm is generally difficult and other topologies, such as the Hilbert norm $\|F - F'\|$, were preferred. It turns out, however, that the trace norm is very convenient for the language of properties and even compulsory when logic is introduced. Microlocal analysis allows a rather simple computation of a trace norm, which is given in a first approximation by the integral of $|f(x,p) - f'(x,p)|/h^n$ on phase space.

Taking into account the fact that all the quasi-projectors have practically the same trace N, one can express the equivalence between the various quasi-projectors by the relation

$$(1/N)\mathrm{Tr}|F - F'| = O(\varepsilon), \quad \text{with } \varepsilon = (\hbar/2\Delta x\,\Delta p)^{1/2}. \tag{10.12}$$

One can then construct true projectors E, E' by straightening the quasi-projectors. The equivalence relations (10.12) remain true for E and E'. They are therefore the equivalence relations we were looking for, and they hold between true projectors and quasi-projectors as well, that is, between the pairs (E, E'), (E, F'), and (F, F').

One can give a simple geometrical interpretation for the indeterminacy in the projectors. The quantity $\mathrm{Tr}|E - E'|$, of the order of $\sqrt{N}$, represents the latitude in the number of unit eigenvalues of E. If the boundary of R is covered by small rectangles with sides σ and σ', the number of these rectangles is also of the order of $\sqrt{N}$. This can therefore be considered as the indeterminacy in the number of semiclassical states associated with R, showing that the result is sensible and cannot be improved.

After straightening the set of quasi-projectors, one obtains a set of projectors that are equivalent and represent the same classical property. Any projector showing this kind of equivalence with the operators in the set can also be considered as belonging to it. One has thus obtained a final answer for the initial question, namely: *a classical property corresponding to sufficiently large a priori error bounds Δx and Δp is represented by a set of equivalent quantum projectors.*

*Note.** Although the logic of physics demands true projectors, the idea of replacing them directly by quasi-projectors is well known in mathematics for similar purposes. It is even a favorite trick in microanalysis. Quasi-projectors should not be therefore considered either as an artificial or an intermediate construction but rather as a perfectly good tool in itself.

MUTUALLY EXCLUSIVE CLASSICAL PROPERTIES

92. The fuzziness we found in the boundary of a classical property raises a question: can one consider as distinct two properties associated with nonoverlapping rectangles? The interest of this question is obvious if one thinks of two measurement data represented by two different positions of a pointer on a dial: up to what point are they distinct phenomena?

The Difficulty

Suppose that a particle is prepared classically in the rectangle R. This corresponds to a density operator $\rho = N^{-1}E$ (as will be shown when the density operator is introduced in Chapter 13). Let E' be a projector associated with the outside of R. The two projectors E and E' are not necessarily obtained from straightening F and $I-F$ with the same quasi-projector F so that E' is not necessarily equal to $I-E$ but only equivalent to it. One can then show that the probability for finding the particle *outside* R is of the order of $N^{-1}\mathrm{Tr}(F-F^2)$, that is, of the order of $\varepsilon = \sqrt{\hbar/2\Delta x\,\Delta p}$. The same is true for two properties associated with two adjacent rectangles, and there is some overlapping between the properties. Their projectors are not orthogonal, and they do not exclude each other completely as two mutually exclusive quantum properties would do. Wherefrom the question we asked.

The difficulty fortunately becomes negligible when two classical properties correspond to nonadjacent rectangles. Let them be represented for definiteness by two identical rectangles R and R' with a relative distance in x, δx between them as in Figure 10.3. Their mutual exclusion when considered as quantum properties would mean that two projectors E and E' respectively associated with them would satisfy $EE' = E'E = 0$, and this is precisely what one finds, to a very high degree of precision. Alhough the difficulty had to be mentioned and will have to be referred to from time to time, it is of no practical significance: *Two clearly distinct classical properties are mutually exclusive when considered as quantum properties.*

One may get a quantitative idea of the quality of mutual exclusion by considering two quasi-projectors F and F' as obtained from Gaussian wave functions with Hepp's method. Straightforward calculations show that the product FF' (or $F'F$) is exponentially small, or more exactly,

$$\|FF'\| = \sqrt{\hbar/\Delta x\,\Delta p}\,O\Big(\exp\big[-\sqrt{\Delta x\,\Delta p/\hbar}\,(\delta x^2/\Delta x^2)\big]\Big). \quad (10.13)$$

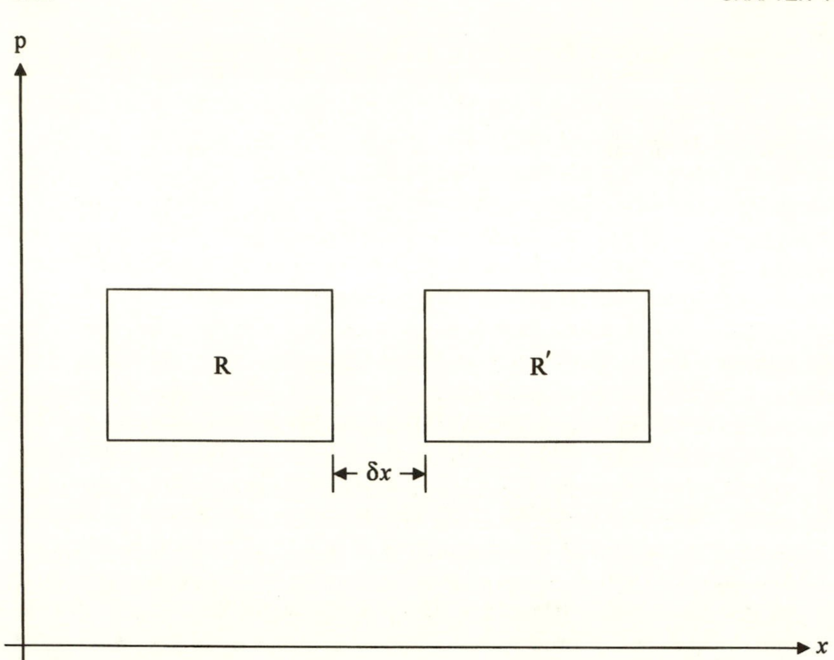

Figure 10.3. Exclusive classical properties. The classical properties associated with nonoverlapping rectangles in phase space are associated with two exclusive projection operators when their distance δx is not too small.

The lefthand side is the ordinary Hilbert norm of the product FF'. Similar results hold for any two projectors E and E', although one can only prove that in place of the exponential in the righthand side of equation (10.13), one has in general a function of the same argument that decreases more rapidly than any power of the previous exponent.

CONCLUSIONS

93. The von Neumann language of quantum properties is *universal* because it embodies also classical properties. Classical dynamical variables and the corresponding properties have a perfectly clear meaning in the quantum language. The description of events occurring during an experiment can therefore be extended to the whole experimental apparatus.

The ideas behind this important conclusion are very simple and the results are quite sensible. They are in full accordance with intuition when the uncertainty relations are taken into account, and this is why von Neumann was already expecting them. He did not succeed, however, in proving his case because of the lack of relevant mathematical techniques at the beginning of the 1930s.

What should one think, then, of the necessity of these elaborate mathematics? Are they an impediment for the clarity of physics? Fortunately, they are not needed in everyday practice, and one can rely on the results, particularly on exclusion, at a qualitative level where excessively small corrections can be entirely neglected.

One may finally add a few words concerning the spirit in which the assimilation between the classical and quantum languages should be made. It would be absurd to allege rigor as a pretext for everywhere replacing commonplace classical statements by their obscure and dense quantum translations. Once we know their fundamental agreement with the quantum principles, they are undoubtedly better with their usual expression. There will still be a few steps on our way toward interpretation where a classical standpoint will have to be proved and its range of validity clearly ascertained, but when this is done, most often once and for all, we can speak classically without further ado.

SUMMARY

A classical property asserts simultaneously the values of position and momentum within error bounds far above the limit of uncertainty relations. It has a meaning as a set of equivalent quantum properties represented by projectors in Hilbert space. Some fuzziness near boundaries does not preclude the mutual exclusion of two nonoverlapping classical properties with extremely good precision.

The language of properties is universal because it can describe the macroscopic features of every apparatus entering into an experiment as well as the microscopic events occurring during the experiment.

11

Classical Dynamics

94. The relation we found in the previous chapter between classical and quantum properties is only a correspondence between two languages. One of them is the ordinary language of classical physics and we might call it classical English for convenience. The other one is specific to quantum physics and the hieroglyphs in which it is written are projectors. It might be called "Copenhagian." We saw previously that Copenhagian can also be expressed in standard English when the sentences it can say are expressed by ones such as "the value of A is in the range Δ," or "this reaction has occurred," and so on. Because of its limitation to the topic of physics and its quantum nature, we may call it "quantum English."

We have thus at our disposal a language, Copenhagian, which is written in its own hieroglyphs and embeds everything classical or quantum-like relevant to physics, which we may wish to express in standard English. This question of language is essential for constructing an interpretation, but it can be considered as already settled.

We turn now in the present chapter to a different question, one that is much more directly akin to the laws of physics. There are two aspects of Copenhagian, corresponding to its reliance on two different principles of quantum mechanics. The translation of a property such as "the value of A is in Δ" relies only on the first principle one gave in Chapter 8. One may notice that this principle was already concerned with language because it stated the mathematical notions that can be used when speaking of a physical system.

When introducing time and using the third principle (the Heisenberg-Schrödinger law of dynamics), we found a simple relation between the projector E expressing "the value of A is in Δ" and $E(t)$ expressing "the value of $A(t)$ is in Δ", namely $E(t) = U^{-1}(t)EU(t)$. Copenhagian is therefore able to express quantum dynamics very simply. This remarkable feature can be attributed to the central role of observables, the projectors being themselves observables, and this is why I insisted in Section 93 on observables as the only real *concept* in quantum physics.

The question I want to investigate in the present chapter is concerned with the effect of this simplicity of dynamics in Copenhagian in the special case of classical properties. It is something deep enough for Bohr to have called it a principle (the correspondence principle). It will not, however, be a new principle because it relies on the previous ones.

What will be its content? Quantum dynamics says that there is a definite one-to-one relation between two projectors such as E and $E(t)$. When these two projectors can express classical properties, there is also a one-to-one relation between the two properties. When these properties are stated in plain English, there will remain an equivalence between them, which is essentially the expression of classical determinism.

The perspective being thus clarified, one can now indicate more exactly the path to be followed in the present chapter. We know that some errors are necessarily involved when a classical property is expressed by a quantum projector. We also had to assume a moment ago that the two projectors E and $E(t)$ represent classical properties, and this is certainly not a universal characteristic. The question to be asked is therefore threefold: (1) Under what conditions is the correspondence between classical properties and projectors maintained through time evolution? (2) What is its expression or, in other words, what is the law of classical dynamics? (3) What errors are involved?

Not so long ago, it would have been impossible to answer these questions, or at least the answers could have been only incomplete and partly conjectural. Rather recent developments in mathematics have completely changed the situation, and the answers are contained in a classical theorem in microlocal analysis by Egorov (1969). The main topic of the present chapter will be a pedestrian version of this theorem for the purpose of physics, with no proof, of course. It will be shown that determinism holds in the majority of cases that are encountered in macroscopic physics. The main exceptions involve chaotic systems or the presence of very narrow potential barriers.

CLASSICAL CELLS AND PROJECTORS

95. The classical properties considered in the previous chapter were very simple. They involved only upper error bounds for position and momentum, and this situation was represented by a rectangle in phase space (at least in the case of a unique degree of freedom, $n = 1$). One needs, however, a more sophisticated class of classical properties when considering dynamics. A rectangular cell cannot be sufficient because it will be deformed under classical motion and take a more complicated shape. Figure 11.1 shows what can happen in the case of a chaotic motion.

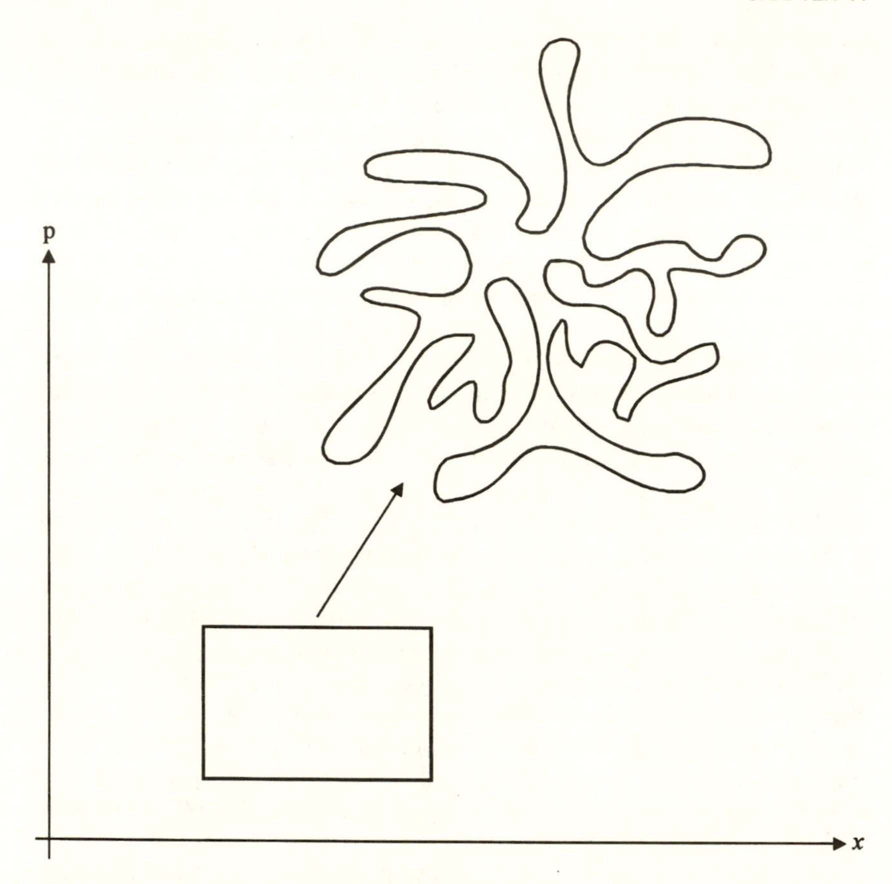

Figure 11.1. An initial cell and its transformation under a chaotic classical motion.

The description of a classical motion must therefore envision domains in phase space more complicated than a simple rectangle. I will concentrate on domains consisting of only one piece, with no hole inside (*connected* and *simply connected*, in the language of mathematics), and they will be called *cells*. A general classical property is therefore expressed by a proposition stating "the point (x, p) is in the cell C."

The relation we found between rectangles and quantum projectors remains practically the same for these more general properties. The methods we introduced previously, using either a sum of Gaussian functions or a microlocal symbol, remain perfectly valid. One can easily guess when the classical-quantum correspondence will break down by looking at the tormented cell in figure 11.1. If the filaments have typical dimensions of the order of Planck's constant, there is not enough room inside for introducing the margin of a microlocal symbol. In the Hepp approach, the width of Gaussian wave functions (in x and also in p after a Fourier transform) are too large for keeping contact with very fine details. Correspondence must be lost in this extreme limit.

On the other hand, one may expect that correspondence holds for good enough cells. I shall call them *regular* and define them by two criteria: (1) the volume of the cell is large in units of Planck's constant, so that they are associated with a large number N of semiclassical quantum states; (2) the cell is rather bulky, that is, "fatty" rather than "twiggy" if these words have a geometrical meaning. This condition must be expressed in more mathematical terms, but its meaning should be clear: the size of the boundary is not very much larger than the inside of the cell.

*Quantitative Aspects**

One can give quantitative criteria instead of these rather vague considerations. It will be enough to consider the existence and the properties of quasi-projectors, from which true projectors can be obtained after straightening. The dimension of the Hilbert subspace associated with the cell C is still given by the number of semiclassical states

$$N=\int_C dq\,dp/(2\pi\hbar)^n. \tag{11.1}$$

A *classicity parameter* ε still controls the equivalence among various projectors so that two of them satisfy

$$N^{-1}\mathrm{Tr}|E-E'|=O(\varepsilon). \tag{11.2}$$

When this parameter is small, the quantum representation of the classical property is correct. When, on the contrary, it becomes comparable with 1, the classical property has no meaning from the standpoint of quantum mechanics.

Everything therefore hinges on the value of ε. To compute it (in the case $n=1$), one can proceed as follows: One introduces a "length" scale L and a momentum scale P, their product being constrained by the condition $LP=2\pi\hbar N$. Also, one introduces nondimensional variables $q'=q/L$ and $p'=p/P$. In a neighborhood of C there is a metric in phase space, which is given by $ds^2=dq'^2+dp'^2$. One can then compute the nondimensional area A of the cell and the length Λ of its boundary (these quantities becoming $2n$- and $(2n-1)$-dimensional integrals when there are n degrees of freedom). The dimensionless ratio $\theta=\Lambda/A$ is a good measure of the geometrical regularity of the cell. It is of the order of 1 for a very bulky cell (a rectangle or a circle) and very large for a tormented chaotic cell.

The scales L and P can then be fixed completely by making θ minimal. With these conventions, the classicity parameter turns out to be

$$\varepsilon = (2\pi\hbar/LP)^{1/2}\theta = \theta/\sqrt{N}. \tag{11.3}$$

One may notice that this parameter enters only in estimates of orders of magnitude, and the factor 2π is therefore an arbitrary convention. It should also be mentioned that the first equality (11.3) is valid whatever the dimension n whereas the second expression involving N holds only for $n = 1$. A simple and exact definition of a regular cell is finally to say that the corresponding classicity parameter is much smaller than 1.

The exclusion property (10.13) can also be extended in the following way: Let C and C' be two nonoverlapping cells. In the previous metric, let a vector with components (L_1, P_1) separate two points in the cells with a minimal distance. One can then write analogously to equation (10.13):

$$\|EE'\| = O(\eta), \quad \text{with } \eta = \varepsilon \exp\left(-(LP/\hbar)^{1/2}\left[(L_1/L)^2 + (P_1/P)^2\right]\right). \tag{11.4}$$

THE CORRESPONDENCE THEOREM

Definition of a Classical Motion

96. There are two principal cases when classical physics is applied. The first is concerned with a particle having a classical motion—in an accelerator for instance, and the Hamiltonian H is then known. More important is the other case, which deals with the manifest, collective behavior of a macroscopic object. Of particular interest then is the collective part H_c of the Hamiltonian as in equation (7.1), which will be also denoted by H, for brevity.

A *Hamilton function* is defined as the symbol $h(q, p)$ of the operator H, itself defined by an equation analogous to equation (10.1). Classical motion is then also defined as satisfying the standard Hamilton equation

$$\dot{q} = \partial h/\partial p, \quad \dot{p} = -\partial h/\partial q. \tag{11.5}$$

One can solve these equations for finding the trajectory of a point in phase space according to classical dynamics.

Classical Dynamics

97. The problem of establishing a dynamical correspondence between quantum and classical physics can then be formulated as follows: Let C_o be a cell in phase space representing a classical property occurring at an

initial time t_o. I will only consider a case when the classical motion between times t_o and t maps C_o into another cell C, which means that the points in C_o move and generate at time t another a cell C (i.e., a set in one piece with no hole). It could happen that this is not true, for instance, when some points starting from C_o are able to cross a potential barrier while other points are reflected. This situation can be obviated by choosing another initial cell every point of which will either cross the barrier or none of them will.

It is important that the two cells C_o and C be regular, that is, large and bulky enough (as a matter of fact, their phase space volume is the same according to Liouville's theorem). This condition is necessary for associating a projector with each of them (or rather a set of equivalent projectors). I will denote by E_o and E two definite projectors that are respectively associated with C_o and C.

One may make a rough guess of the result to be expected by asking what would happen if quantum-classical correspondence were perfect. The projector E would then express the consequence at time t of the property with projector E_o at time t_o. There would then be a projector D evolving in time according to quantum dynamics so that $D(t) \equiv U^{-1}(t)DU(t) = E$ and $D(t_o) = U^{-1}(t_o)DU(t_o) = E_o$, which implies $E = U(t-t_o)E_oU^{-1}(t-t_o)$. Putting $t_o = 0$, the question is whether one has the equality $E = U(t)E_oU^{-1}(t)$ and with what kind of precision. One may remember that E_o and E are defined through the geometrical position and shape of the two cells, which are themselves related by classical motion, whereas $U(t) = \exp(-iHt/\hbar)$ is the Schrödinger evolution operator. When stating the errors, one must not forget that each projector is an arbitrary sample of its own equivalence class.

The question is therefore clearly defined and the work of the mathematician is then to answer it either "yes" or "no" and, when the answer is "yes," to give the errors. A small error will mean that the difference $E - U(t)E_oU^{-1}(t)$ is small, and because this is an operator, some operator topology will be involved.

I will simply state the result without further argument. It can be found in books on microlocal analysis where it is called the Egorov theorem, and it can be adapted for the purpose of physics. The only trouble is that, as usual, a mathematical theorem is never quite general enough for all the applications a physicist would like to cover. The proof of Egorov's theorem is usually given under some restrictive conditions, (assuming, for instance, an infinitely differentiable Hamilton function as well as some bounds on its derivatives). One may presume that some of these conditions can be removed and replaced by weaker ones, and this has been done in some cases, but this kind of situation is familiar in many fields of theoretical physics and we shall not worry about it.

Ergov's theorem. *One has*

$$E = U(t - t_o)E_oU^{-1}(t - t_0), \qquad (11.6)$$

with an error that is essentially the largest classicity parameter for all the intermediate cells through which C_o is going during its classical motion between times t_0 and t before reaching C.

Complements*

For the people wishing to apply the result, a few comments are needed. Since the two cells have the same volume, their number of quantum states N is the same. As usual in microlocal analysis, the errors are best expressed with the help of a trace norm. A more explicit form of Egorov's theorem is

$$N^{-1}\mathrm{Tr}|E - U(t - t_o)E_oU^{-1}(t - t_o)| = O(\zeta). \qquad (11.7)$$

The ζ-parameter controlling correspondence depends on $t - t_o$, on the Hamilton function and its derivatives, and on the cells one is considering. Mathematical papers seldom give an explicit value for this parameter and the job of physicists is to get one. There are more or less complicated expressions for ζ, but there seems to be a simple rule of thumb, which has been included in my statement of the theorem: One considers the cells through which C_o is moving during its motion towards C, at various times t' such that $t_o \leq t' \leq t$, as shown in Figure 11.2. Every such cell has a classicity parameter $\varepsilon(t')$, and one can take an approximation of the dynamical classicity parameter ζ as being the upper bound of the various $\varepsilon(t')$.

This implies a very convenient result: the intermediate cells can be computed from classical physics and their classicity parameters $\varepsilon(t')$ from differential geometry. The value of ζ can therefore be estimated by purely classical calculations. In most cases ζ is small because the cells do not show a tortuous shape down to very small scales, and therefore their classicity parameter remains small.

Correspondence is lost when ζ becomes of the order of 1. The most obvious case of such a violation is a chaotic motion, when chaos reaches a scale of the order of Planck's constant. There are also other cases for which Egorov's theorem does not hold, as can be seen in a detailed proof. They occur when there are very large values of some derivatives of the Hamilton function in some regions through which or near which classical motion is occurring. The only simple physical example I know of is the case of a very narrow potential barrier and it is again rather obvious from the standpoint of physics. There is a large quantum

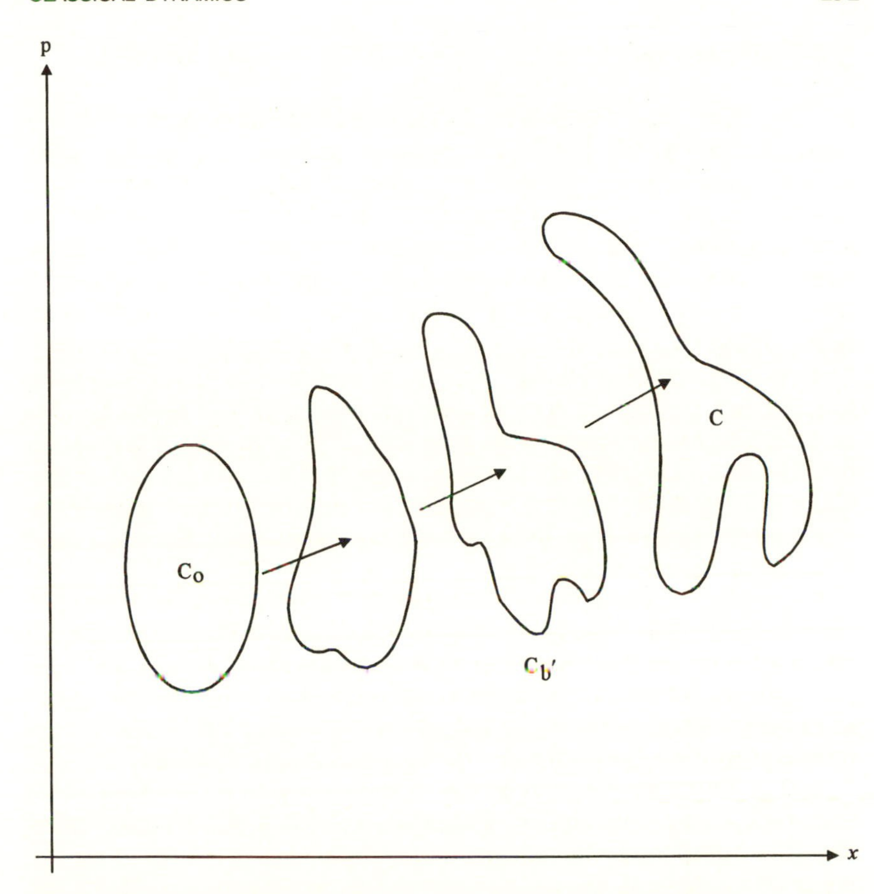

Figure 11.2. The intermediate cells $C_{t'}$ through which the initial cell C_o is transiting before reaching a final cell C. The classicity parameters of these cells control the validity of quantum-classical correspondence (Egorov's theorem).

transmission through such a barrier with little reflection, whereas classical motion always ends up in a reflection when the energy is low enough. Correspondence is therefore a priori excluded.

*Note 1.** To my knowledge, there is no exact proof of the relation of the parameter ζ with the upper bound of the ε's, and it should still be considered as a conjecture when one is dealing with fundamental problems.

Note 2. One may prefer obtaining the errors in a practical case by direct calculations using wave-packet spreading and quantum fluctuations.

APPLICATION: THE STERN-GERLACH EXPERIMENT

98. The experiment that was performed by Otto Stern and his collaborator Walther Gerlach in 1921 remains the most frequently cited quantum measurement. It uses an apparatus in which an atom wave function whose existence is assumed is split into various components according to the value of the z-component of spin S_z, z being a given direction of space. The atom crosses a magnetic field. In the case of a silver atom with a unique electron in an outer S-shell ($l=0$), the atom spin is $1/2$. It remains always in its ground state during the experiment, so that its internal wave function ϕ is unchanging. The total wave function is a *spinor*, which can be expanded as a sum of terms $\Sigma\psi_m(x) \otimes \phi$, m being a spin index with $m = \pm 1/2$. The variable x represents the position of the atom center. In a magnetic field $B(x)$, the Hamiltonian involving the spin and the position but not the internal state of the atom is $H = P^2/2M - gS.B(x)$, M being the total mass, S the spin, and g the gyromagnetic ratio.

The magnet is asymmetric and creates an inhomogeneous field. The atom is sent along the magnet. The field acting on it is directed along a direction z (normal to the trajectory) and depends only practically on the space coordinate z. The Hamiltonian acting on a component $\psi_m(x)$ of the wave function therefore reduces to $H_m = P^2/2M - gmB(z)$. Up to this point, one has used only standard quantum mechanics.

At this point, most textbooks then introduce a mixture of quantum and classical physics essentially as follows: Motion occurs as if the atom, when its spin component in direction z is m, had a potential energy $W(z) = -gmB(z)$. A force $F = -\nabla W$ is therefore acting on the atom. Its motion is essentially classical, and the various m-components follow different trajectories, from which a value for m can be derived when detecting the atom.

This reasoning provides a remarkable example of the versatility in the usual language of physics, showing how intuitive and efficient it can be when compared with the laborious writing and solving of a Schrödinger equation. It is the language we wish to speak without afterthoughts and, in the present case, it is perfectly correct.

*Proof**

If one feels scrupulous about the argument without wishing to enter into tedious calculations with the Schrödinger equation, one can proceed as follows: Let E_m be the projectors in the spin Hilbert space for the two values of m. The Hamiltonian commutes with these projectors and can be written as $\Sigma H_m E_m$, the symbol of H_m being $p^2/2M - gmB(z)$. Egorov's theorem can then be applied to each spin component and this amounts to the previous reasoning.

SUMMARY

In ordinary circumstances, a classical property given at an initial time t_o implies another classical property at a later (or earlier) time t, the two cells in phase space that represent these properties being related by classical motion. This is a result of quantum mechanics, which is valid with known error bounds, which are most often very small. There are exceptions, particularly for a chaotic motion when chaos reaches the scale of Planck's constant and when there are narrow potential barriers. The underlying theorem provides a precise statement of the correspondence principle.

12

Histories

99. The language of properties is universal because it can account for the microscopic features as well as for the classical macroscopic aspects of an experiment. Our intent is now to learn better how to use it.

We have seen how it applies to macroscopic physics. There is no essential difference at this level between the language of properties and the ordinary language of physics, which is plain English in that case. A classical property asserts the position and momentum coordinates of a macroscopic object in the same way and practically with the same words that one uses when speaking of a possible or an observed event. The event occurs at a specific time, and the property catches both the time and the event, like a snapshot. The full course of events, which would be a narrative history of what happens during the experiment, can be unfolded in the same spirit as a series of successive properties, much like a motion picture is a sequence of snapshots. Any classical series of phenomena can be told in this dry manner, whether it be a novel, the history of a famous battle, or the measurements recorded in Michelson's notebooks. As long as one keeps to a macroscopic scale, the language of properties is complete.

The idea of quantum histories consists in using a similar sequence of quantum properties for giving an account of events in the underworld of atoms and particles. In view of what we saw of the relation between quantum and classical properties, we may expect the procedure to work in both worlds, as a universal language. We know, however, that projectors do not commute, and in the past this caused great difficulties with logic. A language of science should be logical, and the consistency of histories will have to be checked in all its aspects.

If a language is logically consistent, every sentence may be considered as a proposition, and the logical game of "and, or, not" and inference can be run with these propositions. The problem is to learn how to play that game when the sentences are quantum properties, that is, noncommuting quantum projectors. Birkhoff and von Neumann showed that this is not trivial, and the range of histories needs to be restricted if they are to make sense.

The present chapter introduces the language of histories. It has a close connection with the ordinary language of physics one hears everyday in a laboratory or during a lecture, and one can start with this simple approach. The question of logical consistency will then appear as soon as one notices that some histories are nonsense for reasons unknown to classical physics, while others seem on the contrary very reasonable. Logical consistency will then lead us to the necessity of probabilities.

INTRODUCING HISTORIES

A Typical Experiment

100. We saw that properties are the main constituent in the language of physics. When the properties describe the course of events in an experiment, they fall naturally under a definite pattern, which is the one of histories. This is illustrated by an example of a more or less arbitrary experiment, for instance in nuclear physics.

The experiment is intended for a study of the reaction $n + p \to d + \gamma$, where a neutron n collides with a proton p to produce a deuteron d and a photon γ, as shown in Figure 12.1. In the ordinary informal language of physics, the course of events in the experiment can be described as follows: Some neutrons exit a nuclear reactor through a channel in the wall of the reactor. A velocity selector lets pass only a few of them with a definite velocity v, up to a small error Δv. The channel is narrow enough and the velocity selection precise enough for the selected neutrons to come out one by one with a rather long time between two successive arrivals. A hydrogen target (i.e., a vessel containing liquid hydrogen) stands behind the selector. When a neutron enters the target, it can interact with a proton in the target, and the reaction $n + p \to d + \gamma$ sometimes occurs. The reaction is detected by a battery of photodetectors all around the target to determine whether a photon is produced.

If the language of properties is used for the same description, one must specify the time at which each event occurs. It will be convenient to distinguish between three kinds of statements entering into the description. The first type gives a classical information concerning the preparation of particles (about the reactor and the exit channel) and other conditions of the experiment (velocity selector, target). We know from the previous chapters that their description by classical physics is in agreement with quantum mechanics. A second kind of statement is concerned with measuring devices (the photodetectors). They are much

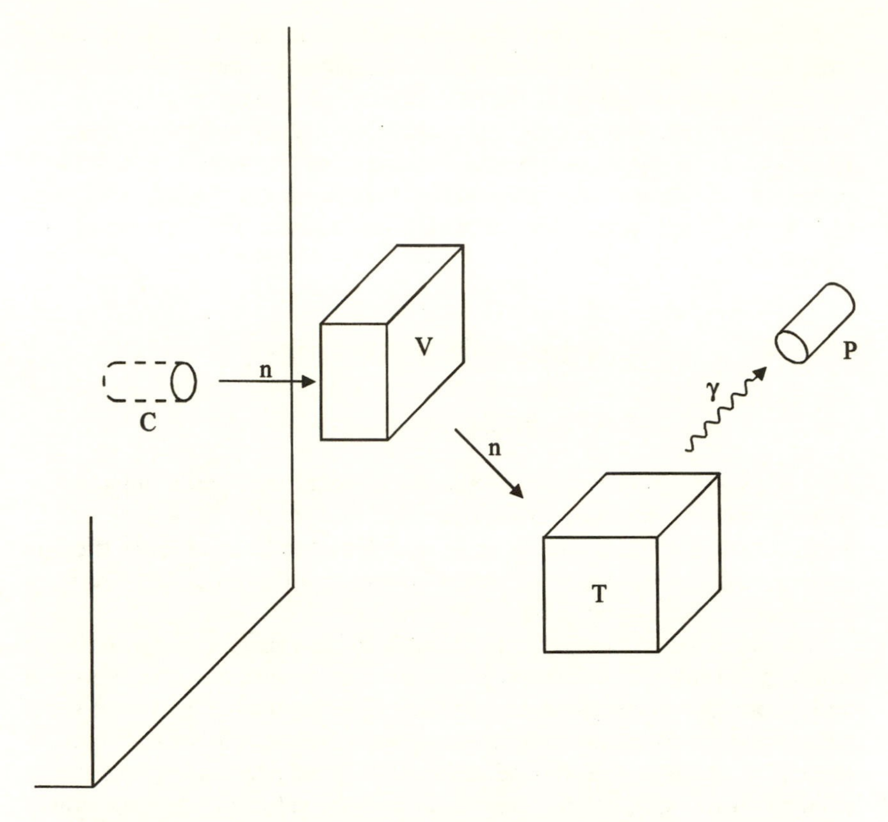

Figure 12.1. An experiment in nuclear physics. A neutron gets out of a reactor through the channel C in the wall of the reactor. It crosses a velocity selector V and then enters a hydrogen target T where a reaction $n + p \to d + \gamma$ takes place. The photon is detected by a photomultiplier P.

less clear, and we shall need the full power of a complete interpretation, which is given near the end of this book, for understanding them. Finally, there are statements about the particles themselves, which belong properly to quantum mechanics. I will only consider this last category for introducing histories.

Taking into account the progression of time, the previous description of the experiment now involves the following sequence of quantum properties:

1. The neutron position is in the exit channel at time t_1.
2. The neutron velocity is in a range Δv around v_o at time t_2.
3. The neutron position is in the target at time t_3.

4. The reaction $n + p \to d + \gamma$ is completed at time t_4.
5. The photon position[1] is in a definite detector (labelled by an index j) at time t_5.

The time ordering of the properties is prescribed, namely

$$t_1 < t_2 < \cdots < t_5. \qquad (12.1)$$

The times $t_1, t_2, \ldots t_5$ are neither perfectly defined nor completely arbitrary. Their choice is constrained by dynamics and also by logic, as we shall see in the next chapters when the consistency conditions selects them. In the present discussion they are taken as given.

Histories

101. Every property we have met is expressed by some projector $E_k(t_k)$, and the description of the experiment amounts to a time-ordered list of projectors:

$$E_1(t_1), E_2(t_2), E_3(t_3), E_4(t_4), E_{5j}(t_5). \qquad (12.2)$$

This set of projectors can be considered as a theoretical construct relying uniquely on the basic principles of quantum theory; however, when seen as a sequence of sentences (of propositions), it is practically a description of a series of events in plain words. Such a sequence of projectors with its descriptive meaning will be called a *history*. The properties it involves must refer to a definite physical system (all the projectors act in the same Hilbert space).

Families of Histories

102. In deterministic classical physics, there is no freedom in the course of events, and a unique history can give a complete and faithful account of the phenomena. In quantum physics, however, there are alternatives because of randomness. Every quantum event could or could not occur. In the previous experiment, a photon could be detected by a detector other than the one with index j. The neutron could miss the target, or it could cross it without interacting. It could also scatter on a proton with no reaction.

One might argue that I did not include randomness among the principles of quantum theory and that it suddenly appears without

[1] Localizing a photon in space requires some care (see the comments at the end of this volume). It is enough for our present purpose to assume that property 5 is effectively represented by a projector and therefore makes sense, even if the construction of this projector is nontrivial.

warrant. This point is worth considering with caution. I did not speak of randomness at the level of principles because it is seen in measurements. A good rule of conduct is certainly to avoid discussing two difficult questions at the same time, and I will not embark on quantum randomness when we should concentrate on the language of physics. We can, however, make room for randomness *as a possibility a language can afford*. Thus, one need not consider a unique history as an account of events endowed with certainty, but rather a whole family of histories representing various possibilities. Any language is basically made of possibilities. It can then easily cover a random situation if it happens to be physical, hence, real. If by chance there were certainty, if histories were only classical and subject to determinism, this would not be a difficulty. I will therefore use the flexibility of language for choosing the most general framework, the one of *possible randomness*. The question of actual randomness will occur later (in Chapter 21) and we shall be ready to deal with it.

Let us consider a *complete* family of histories, meaning that all the different possibilities in the course of events are included. This notion of completeness is not trivial, and one might feel embarrassed if asked the question: have you thought of every possibility? Let us therefore proceed by steps.

A simple method for constructing a complete family of histories was proposed by Griffiths (1984). It consists in fixing once and for all a definite sequence of reference times t_k, the index k running from 1 to n. Also a definite observable A_k for each reference time is specified. The spectrum of A_k is split into a set of domains $\{\Delta_k^{(j)}\}$ that are mutually exclusive ($\Delta_k^{(j)} \cap \Delta_k^{(j')} = \varnothing$ for $j \neq j'$) and complete (the union of all the $\Delta_k^{(j)}$'s covering the spectrum of A_k). The property "the value of $A_k(t_k)$ is in $\Delta_k^{(j)}$" is associated with a projector $E_k^{(j)}(t_k)$. The histories in the family are represented by all the possible sequences $(E_1^{(j_1)}(t_1), E_2^{(j_2)}(t_2), \ldots, E_n^{(j_n)}(t_n))$.

It is clear that all the possible events that can occur under the restrictive conditions of fixed times, fixed observables, and fixed ranges are taken into account, and the family can be considered as complete. Since two histories differ by stating two mutually exclusive properties (with orthogonal projectors) at one time at least, they can be considered as mutually exclusive. In that sense, every possible course of events is included with no redundancy.

Griffiths families are rather restrictive, however, with the straightjacket of their fixed choice of reference times, observables, and ranges. Many experiments require a more flexible description, as was noticed by Gell-Mann and Hartle (1991), and the experiment considered above in nuclear physics provides such an example. That history states that a nuclear reaction is completed at time t_4 and has produced a photon. It is then natural to consider as a possibility that the photon is detected at a later time t_5. The reaction could, however, fail to take place, but this can be anticipated by putting neutron detectors around the target.

One can then consider the following history as a possibility. It coincides with the previous one up to time t_3, then it says that the system is still in the $p+n$ channel at time t_4 and that the neutron is finally detected at a later time t_5'. Because of the difference between the velocity of a photon and a nonrelativistic neutron, the times t_5 and t_5' are certainly different. The two histories refer, therefore, to a common observable at time t_4 (viz. the projector on the *pn* subspace) though with different ranges for this projector (viz. the eigenvalue 0 or 1). The observables and the reference times coming afterward are not any more related: the times t_5 and t_5' are different. The position of a photon and a neutron have nothing in common. Readers will certainly appreciate on their own the wide variety of circumstances one can meet in various experiments and the necessity of using families of histories that are more general than the Griffiths type.[2]

This wide generality raises the questions of knowing when two histories are exclusive and when a family is complete. These are not quite trivial questions, and it will be simpler to obtain a straight mathematical answer. The example of a Griffiths family provides a hint. Its projectors satisfy two kinds of equations:

$$E_k^{(p)}(t_k)E_k^{(q)}(t_k)=0 \quad (p \neq q), \quad \text{and} \tag{12.3}$$

$$\sum_j E_k^{(j)}(t_k)=I. \tag{12.4}$$

Equation (12.3) holds because the two ranges $\Delta_k^{(p)}$ and $\Delta_k^{(q)}$ do not overlap (exclusiveness). Equation (12.4) shows completeness for the properties at time t_k: the ranges $\Delta_k^{(j)}$ cover the spectrum of A_k.

There is already a hint of flexibility in the fact that these relations do not involve explicitly the observables $\{A_k\}$ but directly the projectors. If one denotes the various histories in the family by an index a, one can define a specific operator for each history as the product of its projectors in a decreasing time order, that is,

$$C_a=E_n^{(j_n)}(t_n)\ldots E_2^{(j_2)}(t_2)E_1^{(j_1)}(t_1). \tag{12.5}$$

[2] As a matter of fact, it is (almost?) always possible use Griffiths families, though they may then be somewhat artificial and rather removed from the commonplace language of physics.

The relations (12.3) and (12.4) imply

$$\sum_a C_a = I, \qquad (12.6)$$

where the sum is taken over all histories in the Griffiths family.

More generally, equation (12.6) will be taken as a criterion for the completeness of a family of histories. The histories must also be mutually exclusive. The criterion is that any two histories in the family have at least one common reference time at which they state mutually exclusive properties (with orthogonal projectors). For a wider criterion, see the comments at the end of the book.

HISTORIES AND LOGIC

Some Histories Are Nonsense

103. Histories are as intuitive as quantum physics can tolerate, and they are very close to the standard language of physics. A long time elapsed before they were used because of what looked like an obvious defect: some histories may look very natural, and they are irresistibly tempting from the standpoint of common sense, but they are completely meaningless according to quantum mechanics. In an interference experiment, for instance, one must take into account the various points at which a photon can impinge on a photoelectric screen, or the various photosensitive grains it can interact with in a photographic emulsion. These properties express directly possible observational events. But our mind, our classical mind, is made so that it cannot resist the temptation of asking more and wondering through which slit in a Young device, or through which arm of an interferometer, the photon has gone.

Both kinds of properties can be included in a family of histories. It is a two-time family where each history expresses, at time t_1, which interferometer arm the photon is in, and at time t_2, on which grain of a photographic plate it impinges.

It is well known, however, that this overprecise description is impossible and does not make sense. The reasons are explained in most textbooks on quantum mechanics, and I will not reproduce those arguments here (we shall find better ones later). In any case, one can take for granted that some histories are nonsense. But sense versus nonsense is a matter of logic, and we shall therefore explore this direction of research for improving the language of histories.

Histories and the Operations of Logic

104. The basic question to be now considered is How can one introduce logic in a family of histories?

The family will be assumed to be given. Whether it was introduced for describing an experiment or for investigating a questionable point in interpretation does not matter. Remembering the elements of logic listed in Chapter 8, one can raise the following questions: (1) Can one define a well-defined field of propositions in this family? (2) Can one introduce on this field the three logical operations *not*, *and*, and *or*? (3) Finally, can one define inference?

Let us begin with the first two questions, which are intimately linked, that is, finding the field of propositions and formulating the logical operations. It seems obvious that a property occurring in a history of the family, whatever it may be, should be an elementary proposition. Other, more complicated propositions can be obtained by combining the properties with the help of the words "not," "and," and "or." In the case of a spin-1/2 for instance, an elementary proposition would be "$S_x = 1/2$ at time t_1." But "$S_x = 1/2$ at time t_1 and $S_z = -1/2$ at time t_2" is another proposition. The negation of "$S_x = 1/2$ at time t_1" is the proposition "$S_x = -1/2$ at time t_1" and more generally the negation of a property associated with a projector E is associated with the projector $I - E$.

Rigorous logicians or mathematicians will certainly request more details. They will ask for a precise definition of the logical operations in terms of the projectors or of the quantities $(t_k, A_k, \Delta_k^{(j_k)})$ defining a property in a Griffiths family of histories. It was shown elsewhere (Omnès, 1994a) that one can use the geometric sets $\Delta_k^{(j_k)}$ for representing a history by a set, the direct product $\Delta_1^{(j_1)} \times \Delta_2^{(j_2)} \times \cdots \times \Delta_n^{(j_n)}$ in a n-dimensional space (k going from 1 to n). One can then define a geometric representation of the propositions by all the possible unions of these various sets. The logical operations *and*, *or*, *not* on propositions correspond to the operations *intersection*, *union*, *complement* on sets, which are known to obey the same axioms. This geometric approach does not associate a projector (nor even a definite set of projectors) with a general proposition but only with a property. A history is associated with a sequence of projectors and a general proposition with a set of sequences. This is how one avoids the difficulties that von Neumann encountered in his logical considerations because he assumed that every proposition is necessarily associated with a projector.

Inference and Probabilities

105. To obtain a complete version of logic, the notion of *inference* must now be defined. The trick is to assume that a well-defined probability $p(a)$ can be assigned to every proposition a in the family. If (a, b) is a pair of propositions, the conditional probability for b, given a, is defined by the ratio

$$p(b|a) = p(a \text{ and } b)/p(a),$$

assuming $p(a) \neq 0$. The inference $a \Rightarrow b$ will then be defined by the condition

$$p(b|a) = 1. \tag{12.7}$$

It may be noticed that this criterion excludes the possibility of taking as an antecedent a proposition a with zero probability, but this restriction does not matter, for who would want to draw a consequence from something whose probability of happening is zero?

The essential result justifying this approach is the following theorem:

Theorem. *The logical axioms of inference are automatically satisfied by the definition* (12.7).

These axioms were mentioned in Chapter 8 and the proof of this theorem is straightforward. The essential assumption is that the probabilities $p(a)$ must satisfy exactly the axioms of probability calculus.

The definition of logical equivalence, $a = b$, follows because it amounts to the couple of inferences $a \Rightarrow b$, $b \Rightarrow a$. It can also be written as the two equations

$$p(a \text{ and } b) = p(a) = p(b), \tag{12.8}$$

if these probabilities are nonvanishing.

*Note**

The definition (12.7) is closely related to Bayesian inference, which is used in information theory. There are two other ways of constructing an inference. One of them uses set theory: the propositions are associated with sets and inference is expressed by inclusion. It seems that this approach can lead only to some sort of classical physics (and in any case it applies to strict classical physics). A third method is used in mathematical logic and applied in computers. It states that an inference $a \Rightarrow b$ holds when the proposition *non-*(*a and non-b*) is true. It cannot be applied, however, in quantum mechanics because of the impossibility of assigning a truth value to every proposition, because of complementarity (which will be considered later).

SUMMARY

Quantum dynamics can be used to introduce time in the language of interpretation. Any physical process, and particularly an experiment, can be accordingly described by a history, which is a time-ordered sequence of physical properties. A history is mathematically coded as a sequence of projectors $E_1(t_1), E_2(t_2), \ldots, E_n(t_n)$, the reference times being ordered according to $t_1 < t_2 < \cdots < t_5$.

The language of interpretation allows for the possibility of randomness in a course of events, although it does not demand it. This possibility is accounted by introducing complete families of mutually exclusive histories. Exclusiveness requires that two histories in the family assert mutually exclusive properties (with orthogonal projectors) at least at one common reference time in the two histories. Although completeness is intuitive and simple when the histories are expressed in words, it has an algbraic version: introducing a history operator $C = E_n(t_n) \cdots E_2(t_2)E_1(t_1)$ for each history, completeness amounts to the condition $\Sigma_a C_a = I$ for a sum over all the histories in the family.

13

The State of a System

106. We saw in the previous chapter how probabilities might be used for introducing logic in the language of interpretation. The present chapter is concerned with finding these probabilities for histories. Before any investigation in that direction, however, some points must be made clear. The word "probability" suggests unavoidably randomness and, as was stressed previously, no principle of randomness was introduced in the basic principles of the theory. The probabilities of histories we are looking for cannot therefore express randomness. Their unique purpose at the present stage is to ground the logic of interpretation. They are only numbers associated with histories and satisfying the axioms of probability calculus: mathematical tools for logic and nothing more. Their ultimate relation with a physical randomness will appear only later, when interpretation will be complete enough for obtaining a measurement theory.

With regard to the problem of finding these probabilities, the key point in the present chapter hinges on an obvious statement: the events in a physical process depend on the initial state—the initial situation of the system before anything can be said about it. An attractive answer would be to extend histories back in time for an account of the system's preparation. But as one goes farther in the past, more and more events must be taken into account, and more and more outside objects play a role, so that one would have ultimately to go back to the origin of the universe. One needs, therefore, a convenient summary of past events, which fortunately exists and can be expressed as follows: The initial state of an isolated system is almost never represented by a wave function. Exceptions are met only in particularly simple cases or carefully conducted experiments. Quite generally, however, the initial state can be represented by a *state operator* ρ (also called *density operator* or *density matrix* when the Hilbert space dimension is finite). This operator has three essential properties: (1) It is self-adjoint. (2) It is positive (all its eigenvalues are positive or zero). (3) It has a unit trace ($\mathrm{Tr}\,\rho = 1$).

The probability $p(a)$ of a history a is then given by a formula using the history operator C_a, namely,

$$p(a) = \mathrm{Tr}(C_a \rho C_a^\dagger). \tag{13.1}$$

Most of the later considerations about interpretation will rely on this formula. The fact that it is the only possible one is important, and this issue will not be ignored in the present chapter, except that the reader must be advised that this is mostly a technical problem (a problem of uniqueness) worth only sections with an asterisk warning mark.

Not all the properties of the probabilities (13.1) are dealt with in the present chapter, and their investigation will be pursued in the next one.

A FIRST EXPRESSION FOR PROBABILITIES

The Probability of a Property

107. Even if it was not among the principles of quantum theory listed in this book, we know Born's formula giving the probability of a property when the state of a system is represented by a wave function. I will therefore start from it for inspiration, even if we are aware that something more consistent which relies directly on the first principles will be needed later.

The introduction of Born's formula in textbooks is usually made in two steps and according to the Copenhagen interpretation: (1) The results of a quantum measurement are random events. (2) The probability of a definite measurement result (which can always be expressed by a property) is given by Born's formula. This, however, is not sufficient for my purpose. Most of the relevant properties in histories do not result directly from a measurement. Moreover, we are not yet able to introduce the random character of quantum events, and the probabilities we need are intended only for the purpose of logic.

Yet nothing forbids our taking inspiration from known empirical formulas. The probability for obtaining a (nondegenerate) eigenvalue a for an observable A as the result of a measurement made at time t is given according to Born by $p = |\langle a|\psi(t)\rangle|^2$, where ψ is a wave function one assumes to exist. Introducing the projector $E = |a\rangle\langle a|$ and taking into account the evolution of the wave function $|\psi(t)\rangle = U(t)|\psi\rangle$, this is equivalent to

$$p(a) = \langle \psi|E(t)\psi\rangle = \| E(t)\psi\|^2, \tag{13.2}$$

a formula that is easily extended to cases when there is degeneracy and/or a continuous spectrum.

The Case of Position

As an example of the continuous case, one can consider the property of a particle in one-dimensional space according to which the value of the position observable X is in the range Δ at time 0. The corresponding projector is $E = \int_\Delta |x\rangle\langle x| dx$ and one has $\psi(x) = \langle x|\psi\rangle$ so that equation (4.7) becomes

$$p = \int_\Delta |\psi(x)|^2 \, dx = \int_\Delta \langle\psi|x\rangle\langle x|\psi\rangle \, dx = \langle\psi|E\psi\rangle.$$

In view of $E^2 = E$, this can be written as

$$\langle\psi|E^2\psi\rangle = \langle E\psi|E\psi\rangle = \|E\psi\|^2.$$

Note. The time dependence results from

$$\langle\psi|E(t)\psi\rangle = \langle\psi|U^{-1}(t)EU(t)\psi\rangle = \langle\psi(t)|E\psi(t)\rangle.$$

The extension of equation (13.2) to the probability of a history is far from obvious and we shall first try a guess to get some insight, before trying to justify it. Consider the history operator C_a given in equation (12.5) and try

$$p(a) = \|C_a\psi\|^2 = \langle\psi|C_a^\dagger C_a\psi\rangle. \tag{13.3}$$

Introducing the one-dimensional projector $E_o = |\psi\rangle\langle\psi|$, this quantity is also

$$p(a) = \mathrm{Tr}(C_a E_o C_a^\dagger). \tag{13.4}$$

The second equation is obtained by using the cyclic invariance of a trace, in view of which the first formula can be written as $\mathrm{Tr}(C_a^\dagger C_a E_o)$. The trace is computed by using an orthonormal basis in Hilbert space whose first vector is $|\psi\rangle$, from which one obtains equation (13.4).

108. The justification of formula (13.3) can be approached in various ways. When all the properties occurring in the history result from a measurement, this formula gives the joint probability of all the measurement results, if one accepts the Copenhagen rules. This was shown by Aharonov, Bergman, and Lebowitz (1964). We cannot, however, rely on their approach for three reasons: because no measurement is generally at the origin of a history; because we do not know whether wave functions exist; and finally, because wave function reduction enters into their proof and its meaning must be considered as especially questionable.

One might use handwaving arguments and say that equation (13.2) means that the projector $E(t)$ acts on the wave function ψ by conserving the projection of ψ on the eigenvectors of $E(t)$ with eigenvalue 1 and suppressing the other components. Iteration of the process suggests that the wave function $C_a\psi$ is the outcome of a history, but that is a guess.

The most convincing intuitive justification is obtained when every observable A_k in the history is a position or a momentum (one or the other for each reference time). The action of the operator C_a on a wave function is then found to select the Feynman paths in phase space crossing the different "windows" $\Delta_k^{(j_k)}$ at the different times t_k. This looks like a reasonable version of a history. The quantity $C_a\psi$ is considered as a probability amplitude, from which the probability (13.3) follows. Although not a rigorous proof, it looks at least sensible and sufficient for a start. I will therefore rely on equation (13.3) before coming back to the question of its proof.

STATE OPERATOR AND PREPARATION

109. According to Dirac, the initial state of a quantum system is determined by its preparation process. One can rely on this idea to find how a state should be described. This argument will again be a physical one rather than a formal proof, which will be given later. The general trend of this argument is as follows: (1) Together with equation (13.4) for the probability of a history, one assumes the existence of a wave function for describing the initial state of any system entering in the experiment. (2) One makes explicit the fact that a preparation device must be itself prepared, and this fact is taken into account by extending the history we have toward the past so as to include preparation. (3) The existence of a wave function is then found to be inconsistent, and it must be replaced by a state operator. (4) Equation (13.4) for the probability of a history must then be replaced by equation (13.1), which involves the state operator.

Let one consider as an example an experiment in nuclear physics involving the reaction $p + d \rightarrow He_3 + \gamma$. The experiment is described by a family of histories among which one selects a special one, a, associated with a sequence of projectors $(E_1(t_1), \ldots, E_n(t_n))$. The proton is prepared by an accelerator and the reaction occurs in a deuterium target. One begins by assuming that the state of the whole system, including the free incoming proton and every experimental device with the exception of the accelerator, is given by a wave function ψ at a time t_o. This time is prior to all the events in the history and later than the exit of the proton from the accelerator. The assumed existence of the wave function can be expressed by a projector $E_o(t_o) = |\psi\rangle\langle\psi| = B$, where the notation B stands for "beginning." Using the history operator $C_a = E_n(t_n) \ldots E_2(t_2)E_1(t_1)$, the probability (13.4) of history a is given by

$$p(a) = \mathrm{Tr}_S(C_a B C_a^\dagger),$$

where S denotes the whole experimental system including the proton (but not the accelerator). One may notice that B represents an input, an initial information which does not properly belong to the history because its negation has no sensible meaning (one does not see what meaning could be attributed to a statement saying that the initial wave function is not ψ).

Let us now extend the history by including in it the preparation process. Using the same procedure and assuming again the existence of an initial wave function, one must now introduce a larger initial wave function Ψ for the system consisting of the accelerator, the proton in it, and the rest of the experimental devices. The new initial time is denoted by t_i, which is earlier than t_o. One must also use a larger history that includes the working of the accelerator as well as the events that were already included in the first history a. One might say for instance that the proton is injected in the accelerator at a time t_{-2} and is ejected at time t_{-1}. Every such property is described by a projector. In place of relying on the proton wave function and the beginning projector B at time $t_o > t_{-1}$, one has now a history b that is associated with the operator $C_b = E_n(t_n) \ldots E_2(t_2) E_1(t_1) E_{-1}(t_{-1}) E_{-2}(t_{-2})$, which can also be written as $C_a C_i$ with $C_i = E_{-1}(t_{-1}) E_{-2}(t_{-2})$.

The new beginning projector is $B' = |\Psi(t_i)\rangle\langle\Psi(t_i)|$ and the probability of history b is

$$\mathrm{Tr}_{A+S}(C_a C_i B' C_i^\dagger C_a^\dagger),$$

where A denotes the accelerator.

The previous history a continues to make sense although it has now become a part of the larger history b. Its probability can still be obtained, although as a partial series of events belonging to the larger series b. This extraction of an event from its conditioning is expressed in probability theory by a conditional probability, that is, the probability of the history symbolized by C_a given the preparation symbolized by C_i. This conditional probability $p(b|i)$ is therefore given by

$$p(a) = \mathrm{Tr}_{A+S}(C_a C_i B' C_i^\dagger C_a^\dagger) / \mathrm{Tr}_{A+S}(C_i B' C_i^\dagger).$$

A simpler expression can be obtained if one performs the trace on A first, leaving in place of C_i and B' an operator in the Hilbert space of S that is given by

$$\rho_S = \mathrm{Tr}_A(C_i B' C_i^\dagger) / \mathrm{Tr}_{A+S}(C_i B' C_i^\dagger). \qquad (13.5)$$

The probability $p(a)$ is then given in terms of this operator by

$$p(a) = \mathrm{Tr}_S(C_a \rho_S C_a^\dagger). \tag{13.6}$$

When the preparation process is taken into account and one conceives of the macroscopic devices as obeying quantum mechanics (as required by our universal language), one therefore finds that probabilities do not rely on the existence of a wave function. The initial state is described by the operator ρ_S, which will be denoted more simply by ρ. The following properties of this state operator are easily derived from its expression (13.5):

$$\rho = \rho^\dagger, \tag{13.7}$$

$$\rho \geq 0, \tag{13.8}$$

$$\mathrm{Tr}\,\rho = 1. \tag{13.9}$$

(The first and third properties are obvious, the second one is a consequence of equation (13.5) and the positivity of B').

One might still go back earlier in the past and consider the proton injector and—why not?—the materials that were used several years before for constructing the accelerator ("preparing" it). Some people do not waver at considering an initial wave function of the universe. Fortunately, one does not need to be so daring and one can rely on a simple practical rule: *the initial state of an isolated physical system can always be described by a state operator* ρ *satisfying the conditions* (13.7–9), *the probability of a history being given by equation* (13.6).

One may conclude this important result with a few remarks:

1. When a system Q belongs to a larger system S in which it is (momentarily) isolated, its state operator is given by

$$\rho_Q = \mathrm{Tr}_{\bar{Q}}\, \rho_S, \tag{13.10}$$

where the trace is taken over all the degrees of freedom of S not belonging to Q.

2. The probability of a property with projector $E(t)$ (i.e., a one-time history) is given by

$$p = \mathrm{Tr}(\rho E(t)). \tag{13.11}$$

3. Although equation (13.6) is the only possible expression for the probability of a history, it is not claimed that it satisfies all the axioms of probability calculus when inserted in a complete family of histories. This important issue will be considered in the next chapter.

Pure States and Mixtures

110. Equation (13.11) entails a simple definition of the state: the initial state of a system is a summary of its preparation that gives, in principle, the probability of any property. It can also give the probability of a history (subject to consistency conditions that will be described in the next chapter). It is expressed mathematically by a state operator.

By a convention owing to the history of quantum mechanics, a state is said to be pure when one can use a wave function. One can do so when the state operator has an eigenvalue 1 (all other eigenvalues being zero in view of equation (13.9)). The corresponding eigenvector is called the wave function. Otherwise, the state is said to be a mixture.

Although the existence of a wave function is exceptional, there are still strong reasons for using it as much as possible. First, wave functions are very rare in nature, but not so uncommon in a laboratory, where one tries as far as possible to work in perfectly defined conditions. A second reason is the relative triviality of the difference between using a wave function or a state operator in calculations. One can reduce the state operator to a diagonal form $\rho = \Sigma p_j |j\rangle\langle j|$ and then treat each eigenvector $|j\rangle$ like a wave function. Since the most difficult and the most interesting part of a calculation is to find what happens to this wave function and what can be learned about it, going back to ρ at the end of the calculation is more or less trivial and can be often left aside. This is not always the case, however, and the use of a state operator is essential in statistical physics. It is as essential in interpretation, also, and this is why one will rely on it from now on.

OTHER APPROACHES TO THE STATE OPERATOR*

I intend now to add some complements and especially to return to the various questions of rigor that were left aside in the previous intuitive approach to the state operator. The next three sections are accordingly somewhat technical and they can be omitted in a first reading. As an alternative, the reader is invited to read Section 114 on the practical description of an experiment, which is much closer to real physics.

The Conventional Approach*

111. Lev Landau, Felix Bloch, and von Neumann independently introduced the state operator by assuming the existence of a wave function but allowing for a poor knowledge of this function. They assumed that an ideal preparation process would define a specific wave function ψ_j.

The preparing apparatus itself can be imperfect, however (for instance, because of random (classical) fluctuations during its operation). Our inability to control these details implies a randomness of the wave function itself, according to a classical notion of probability expressing ignorance. The result is a description of the state by an operator.

One can assume that there exists a set of possible wave functions ψ_j, each one of them having a probability p_j of being produced (they are not supposed to be mutually orthogonal). The probability of a property a with a projector E for a wave function ψ_j, is given by the Born formula $\|E\psi_j\|^2$. Taking into account a probability p_j for producing the wave function ψ_j, the probability of property a is given by the rule of compound probabilities as

$$p(a) = \sum_j p_j \|E\psi_j\|^2.$$

One can easily show that this is identical with equation (13.11) (with $t = O$) because $p(a) = \mathrm{Tr}\rho E$ with $\rho = \Sigma_j p_j |\psi_j\rangle\langle\psi_j|$. The properties (13.7–9) follow immediately: Equation (13.7) results from the self-adjointness of the various projectors $|\psi_j\rangle\langle\psi_j|$ and the reality of the p_j's. The inequality (13.8) results from $\langle\phi|\rho|\phi\rangle = \Sigma_j p_j |\langle\phi|\psi_j\rangle|^2 \geq 0$, for any vector ϕ. The normalization conditions (13.9) results from $\Sigma_j p_j = 1$.

Gleason's Method: The Case of Properties*

112. Gleason's method for introducing the state operator is the most abstract one. It is based on a mathematical theorem asserting the necessary existence of this operator if a probability is to exist for every property. This is a typical existence theorem that gives no hint about the explicit construction of the existing object. The method is nevertheless useful because it leads to the existence and uniqueness we are looking for.

The method relies on a few axioms. Introducing a probability $p(E)$ for every property, one assumes that $p(E)$ depends only on the property, that is, on E itself and nothing more. This means that if E projects on a subspace M in Hilbert space, the probability does not depend on a special choice of basis in M. One assumes furthermore that $p(E) \geq 0$ and $p(I) = 1$.

Then one considers mutually exclusive properties, corresponding, for instance, to a definite observable A and two nonintersecting sets Δ_1 and Δ_2 in the spectrum of A. The nonoverlap of the two ranges implies that the two properties are mutually exclusive, if the meaning of the statement "the value of A is in ..." is to make sense. The associated projectors E_1 and E_2 satisfy the conditions $E_1 E_2 = E_2 E_1 = 0$. This is, in fact, a general expression of mutual exclusion because, when two projectors satisfy those conditions, there always exists an observable A and two nonintersecting ranges Δ_1 and Δ_2 generating these projectors. Noticing that the property with range $\Delta_1 \cup \Delta_2$ is associated with the projector $E_1 + E_2$, one can assume the additivity property for probabilities to be given by

$$p\left(\sum_j E_j\right) = \sum_j p(E_j) \tag{13.12}$$

for any finite or countably infinite collection of projectors $\{E_j\}$ such that $E_j E_k = E_k E_j = 0$ when $j \neq k$.

Gleason added two more technical assumptions. The first one requires a separable Hilbert space (i.e., having a finite or a countable basis). The dimension of the Hilbert space, furthermore, is larger than 2.

> The case of dimension 1 is trivial because there is only one projector and equation (13.12) does not make sense. For dimension 2, one can think of a spin 1/2, and every projector can be associated with a unit vector n in the three-dimensional Euclidean space by a relation $E = (1/2)(I + \sigma.n)$ into which the Pauli matrices enter. The probability is therefore simply a function $p(n)$. Given a projector E, there is only one projector with which it is exclusive, namely $I - E$. The condition (13.12) reduces therefore to $p(n) + p(-n) = 1$, a condition so weak that it leaves a large freedom for $p(n)$. In practice, however, the real world is made of particles represented by infinite-dimensional Hilbert spaces, and a low-dimensional Hilbert model must be the result of drastic simplifications. The probabilities to be used in such a model should be understood as probabilities of a property of real particles under special conditions. This is particularly true of a pure spin 1/2, which is only a theorist's toy. In other words, one can ignore the restriction on the Hilbert space dimension when applying Gleason's theorem to physics.

Gleason's theorem states that under these conditions, the possible expression of $p(E)$ is strongly restricted: it has necessarily the form of equation (13.12) in which a state operator ρ obeying equations (13. 7–9) must enter. This basic theorem implies a couple of important consequences: The existence of a state operator is necessary for the existence of probabilities. The probability of a property has a unique expression.

The Case of Histories*

113. Gleason's method can also be used for showing the uniqueness of equation (13.1) for the probability of a history. The proof is technical, and I shall only mention the assumptions and the restrictions entering in it.

The restrictions concerning the underlying Hilbert space are the same as in Gleason's theorem. Some further assumptions are directly related with the logical role of histories in a sensible language, and they rely on the idea of tautology, which comes from logic. One says that there is a tautology in a history when it states the same property twice with no intermediary. This is tantamount to saying "my sister is here" and immediately stating, "you know, she's here." This happens, for instance, when the property with projector $E_{k+1}(t_{k+1})$ repeats essentially what is already stated by the previous property $E_k(t_k)$, that is, $E_{k+1}(t_{k+1})E_k(t_k) = E_k(t_k)$. In that case one assumes that the second property can be deleted from the account of events (doing so gives a shorter history with $n-1$ reference times). Similarly, one may consider a first property $E_1(t_1)$ in a history as a tautology when it states something that is already contained in the initial state operator so that $E_1(t_1)\rho = \rho$.

Equation (13.1) for the probability of a history and the uniqueness of this formula result from these assumptions. The proof is, however, limited by further assumptions: (1) One may restrict the number of reference times to $n = 2$ (the case $= 1$ being Gleason's theorem), and the uniqueness of equation (13.1) follows. (2) One can restrict all the properties in the histories to those concerned with position or momentum observables. Using Feynman path integrals, the result is then obtained for any n. One may guess, however, that the result is universal and has not yet been proved because of an insufficient exertion.

Note added in proof: A general proof of uniqueness, relying essentially on the same assumptions and valid for any n, has been given recently by Giuseppe Nisticò (*Assigning probabilities to quantum histories*, preprint 10/98, Department of Mathematics, University of Calabria, Arca Vacata di Rende, CS, Italy).

THE DESCRIPTION OF AN EXPERIMENT

114. I now turn to much more practical aspects of interpretation. The purpose of the present section is twofold: It will indicate in more detail how the description of an experiment is expressed in the language of

histories. It will also show how one can justify some standard models that are frequently used in the practice of physics.

Let us first recall that an experimental apparatus can involve several macroscopic devices for preparing, conditioning, and measuring some quantum objects, which we also want to take into account. I will leave aside the preparing apparatus, whose action can be replaced by giving the initial state. I will also ignore temporarily the measuring devices, whose discussion requires a fully elaborated interpretation.

Let S denote the microscopic system (atoms, particles, and so on) of interest. In order to describe the macroscopic apparatus A, one must introduce some relevant variables giving its location, shape, composition, and operation. When this is done, it is convenient from the standpoint of theory, to consider the apparatus as made of two interacting systems as in Section 58: a collective system (denoted by c) and an environment e representing all the internal degrees of freedom. The trace in equation (13.1) that gives the probability of a history must then be performed on the whole system $S + c + e$.

Before going further, an important warning should be repeated: *Like any significant language, the language of histories is intended to make sense of the real world. It does not pretend, however, to include every aspect of reality but only what is relevant for understanding it. This is contrasted with the language of mathematics, which can be more complete.*

This means that the histories describing an experiment will include only some properties that are *relevant* for *understanding* a physical process. The relevant properties are of two kinds, either classical or quantum. The first category involves the description and working of the apparatus, and it is best expressed in classical terms, even though we know it can be given a quantum meaning. These classical properties are concerned with the collective part, c, of the apparatus.

The environment is essentially defined as something one ignores and to which there is no empirical access. It can be described at best by statistical physics when one needs global information about it, for instance, temperature (i.e., the average internal energy).

The second category of properties is concerned with what happens at a microscopic level and is relevant for understanding physical processes. They are concerned with the S system.

When different possible courses of events are taken into account, one is left with a complete family of histories, and one can calculate the probability of each history. Our intent is now to see what kind of physical insight we may get from equation (13.1), which gives this probability.

The trace in equation (13.1) is over the entire system, i.e., $c + e + S$. It is more convenient to compute it in three steps, namely performing first the trace on e, then on c, and finally on S. The role of the environment was traditionally considered as unimportant (could something important come from pure ignorance?), but everything has changed with the recognition of decoherence. The effects resulting from the trace on e will be described later when we will consider the decoherence effect. For the time being, we may suppose that this trace has been made, with the unique though essential consequence that we do not have to worry about macroscopic interference effects for collective properties.

Then the trace over the collective degrees of freedom, c, is performed. For appreciating its effect, it will be convenient to write more explicitly the history operator in equation (13.1) after separating property projectors from the time-evolution operators. Taking as time zero the initial time at which the initial state operator (of $c + S$) is given, one has

$$C_a = U(t_n)^{-1} E_n U(t_n - t_{n-1}) \dots E_2 U(t_2 - t_1) E_1 U(t_1). \quad (13.13)$$

The evolution operator $U(t)$ is given by $\exp(-iHt/\hbar)$, where H is the Hamiltonian for the system $c + S$ (including the interaction of its two parts). The various projectors represent quantum properties of S as well as classical relevant properties of the apparatus, which are described quantum-mechanically.

*Note.**

The study of decoherence in Chapter 18 will show how the environment can affect the Hamiltonian of c after taking the trace on e. Typical of this are pressure effects, but this consideration is not essential for our present purpose and will be ignored.

When the trace on c is performed, it can sometimes produce useless results. This happens when the quantum properties one tried to introduce were not relevant for the understanding of physics. An explicit calculation of this partial trace, furthermore is, often impossible, except in favorable cases.

Under normal conditions, however, one is able to extract the essentials of the trace on the collective degrees of freedom because it has a clear intuitive meaning. We shall only consider two of these essential aspects because otherwise all the insights and the strategies of physics would have to be reviewed.

The working of the apparatus often introduces time-dependent macroscopic conditions (such as a time-varying field, for instance), which are reflected on S by a time evolution $U(t, t')$ depending on both t and t' (rather than a time-translation invariant $U(t - t')$. The U we are now considering is an operator in S-space—it is not necessarily expressed as a function of a time-dependent S-Hamiltonian nor is it always unitary, though everything is much simpler when it is.

Another typical effect of the trace on c is to give boundary conditions for the S-Schrödinger equation. In a two-slit interference experiment, for instance, there is a complicated interaction between a photon and the matter composing the screen. The exact result of the trace on c would be extremely complex if it were written down in full detail. It is, however, reasonable to replace it by geometric boundary conditions, and doing so can even be justified by simple models from which one can obtain error estimates, (which are most often irrelevant).

The above remarks may probably be of some help for people who want to see through the formalism of quantum mechanics more clearly. Their outcome is commonplace since it boils down to what is done everyday in physics. The justification, as we saw, is not so trivial, however, and the mathematical aspects of the history language may be required for transforming old habits and conventions into well-defined notions.

SUMMARY

The preparation of a physical system can be summarized by an initial state operator ρ, which is a self-adjoint positive operator with a unit trace. It can sometimes be reduced to a wave function.

A history a describing later physical processes can be associated with a probability $p(a)$ for logical purposes. This probability is given in terms of the history operator C_a by $p(a) = \mathrm{Tr}(C_a \rho C_a^\dagger)$.

14

Consistent Histories

115. Our present search for a language of interpretation is reaching an end. We presumed the language of histories to be universal. We found a logical framework for it, but it demands well-defined probabilities for histories. We have found the expression of these probabilities. So what remains to be done?

We must make sure we have obtained a genuine probability that satisfies the three basic axioms of probability calculus. This is a perfectly well-defined problem, but the answer can be inferred beforehand: the most demanding axiom is additivity, which says that the probability for the union ("or") of two exclusive events (histories) is the sum of their probabilities. But quantum mechanics is a linear theory in which amplitudes are additive, and they do not easily fit the additivity condition for probabilities. This obstacle is the possibility of interferences, and it must be ruled out.

Additivity is equivalent to a set of algebraic equations, or *consistency conditions*, that were found by Robert Griffiths (1984). They stand at the acme of the correspondence between language, intuition, and the formalism of quantum mechanics—at the junction of practical physics and symbolic physics.

The Griffiths consistency conditions provide a necessary and sufficient criterion for the validity of logic. They are most often satisfied for two main reasons: either the histories rely (at least partly) on valid classical assumptions, or the decoherence effect is responsible for suppressing some pernicious interferences. Because of the importance of the decoherence effect, Gell-Mann and Hartle (GMH) proposed a criterion stronger than the Griffiths conditions for having additivity and therefore logic (1991). This GMH condition is a sufficient condition and not a necessary one, but acquaintance has shown that it is significantly simpler than the Griffiths conditions and I will use it systematically.

The conclusions of this chapter mark a turning point in our construction of interpretation. The consistency criteria will allow a sound use of logic. With it, we shall have at our disposal a universal language for describing physical processes, for reasoning about them, and for logically drawing the consequences of observations. Interpretation will be a deductive theory.

CONSISTENCY CONDITIONS

116. Among the three axioms of probability calculus (positivity, normalization, and additivity), only the first one is obviously satisfied by formula (13.1) for the probability of histories.

A cyclic permutation of the trace in equation (13.1) can be used to write the probability $p(a)$ as $\mathrm{Tr}(\rho C_a^\dagger C_a)$, that is, the average value of a positive operator $C_a^\dagger C_a$.

In order to investigate the problem of additivity, it will be convenient to work with a simple case as an example. It consists of a two-time family of histories into which two properties can enter at each time. More precisely, the two times are denoted by t_j ($j = 1, 2$). At each time, one considers two *dichotomic* or "yes-or-no" properties. One of them states, for instance, "$A_j(t_j)$ is in Δ_j" and the other one "$A_j(t_j)$ is in $\overline{\Delta}_j$," $\overline{\Delta}_j$ being the complement of Δ_j in the spectrum of A_j. The associated projectors are denoted by E_j and $\overline{E}_j$, and they satisfy the relation $E_j + \overline{E}_j = I$.

It will be convenient to represent the histories in a two-dimensional diagram (Figure 14.1) where the two axes are the spectra of A_1 and A_2. The history "$A_1(t_1)$ is in Δ_1 and $A_2(t_2)$ is in Δ_2" is represented by a rectangle with sides Δ_1 and Δ_2 (i.e., the direct product $\Delta_1 \times \Delta_2$). There are four histories and the logical operations *and*, *or*, and *not* can act on

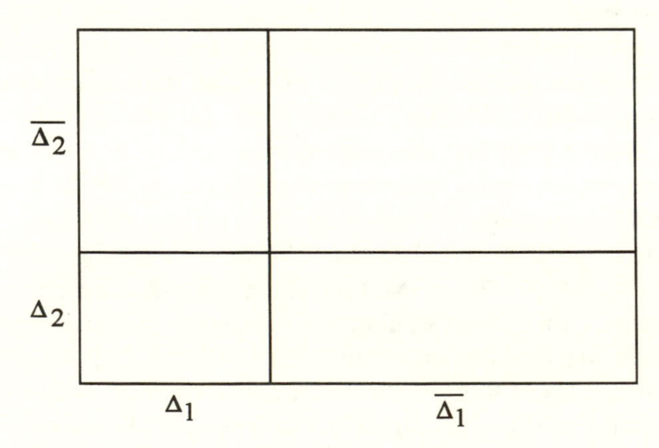

Figure 14.1. Graphical representation of a simple family of histories.

them like the set operations *intersection*, *union*, and *complement* on sets. The field of propositions consists of all the subsets of this set of four elements, that is, it involves $2^4 = 16$ propositions (including the trivial proposition associated with the whole set and the impossible one associated with the empty set).

This geometric approach may look rather abstract, and some readers may prefer a more intuitive example with more physical content. But this is precisely what we must avoid, making sure that no classical habits of thought that are unjustified from the standpoint of principles will be used in our reasoning.

117. We shall investigate the additivity of probabilities in the case of the two histories $a = (E_1, E_2)$ and $b = (\overline{E}_1, E_2)$ from which the time reference has been omitted for brevity. The probabilities are represented by the two rectangles $\Delta_1 \times \Delta_2$ and $\overline{\Delta}_1 \times \Delta_2$ whose intersection is empty. These histories are therefore mutually exclusive. The proposition $c = a$ or b is represented by the rectangle $\Sigma_1 \times \Delta_2$ (where Σ_1 is the spectrum of A_1); it is also a history with projectors (I, E_2). Additivity implies in that case the equation $p(c) = p(a) + p(b)$. When the three probabilities are written as traces with the help of equation (13.1), this condition becomes

$$\operatorname{Re}\operatorname{Tr}\left[E_1(t_1)\rho\overline{E}_1(t_1)E_2(t_2)\right] = 0, \tag{14.1}$$

where the reference times have been reintroduced for later use, and the notation Re denotes as usual the real part of a complex number.

*Proof**

One has $p(a) = \operatorname{Tr}(E_2 E_1 \rho E_1 E_2)$, which can be written after a cyclic permutation in the trace as $\operatorname{Tr}(E_1 \rho E_1 E_2^2)$ or $\operatorname{Tr}(E_1 \rho E_1 E_2)$ by using the relation $E_2^2 = E_2$. Similarly, $p(b) = \operatorname{Tr}(\overline{E}_1 \rho \overline{E}_1 E_2)$. It is convenient to write $p(c)$ as $\operatorname{Tr}(I\rho I E_2) = \operatorname{Tr}[(E_1 + \overline{E}_1)\rho(E + \overline{E}_1)E_2]$. The relation $p(c) = p(a) + p(b)$ implies thereby

$$\operatorname{Tr}(E_1 \rho \overline{E}_1 E_2) + \operatorname{Tr}(\overline{E}_1 \rho E_1 E_2) = 0.$$

The two traces in this equation are complex conjugate since one has $\operatorname{Tr}(E_1 \rho \overline{E}_1 E_2)^* = \operatorname{Tr}(E_1 \rho \overline{E}_1 E_2)^\dagger$ and in view of the hermiticity of all the operators, one has $(E_1 \rho \overline{E}_1 E_2)^\dagger = E_2 \overline{E}_1 \rho E_1$ so that $\operatorname{Tr}(E_1 \rho \overline{E}_1 E_2)^* = \operatorname{Tr}(E_2 \overline{E}_1 \rho E_1) = \operatorname{Tr}(\overline{E}_1 \rho E_1 E_2)$, the last equality resulting from a cyclic permutation. The additivity condition is therefore $\operatorname{Tr}(E_1 \rho \overline{E}_1 E_2) + \operatorname{Tr}(E_1 \rho \overline{E}_1 E_2)^* = 0$, which is equation (14.1).

One can obtain another consistency condition from every pair of mutually exclusive histories a and b such that "a or b" is another history. In the present case, however, they are either trivial or reduce to Eq.(14.1) so that there is only one consistency condition.

Proof *

The case $a = (E_1, E_2)$, $b = (E_1, \overline{E}_2)$ (or $a = (\overline{E}_1, E_2)$, $b = (\overline{E}_1, \overline{E}_2)$) is trivial. Replacing ρ in the first case by $\sigma = E_1 \rho \overline{E}_1$, additivity for the two histories reduces to additivity for two properties with projectors E_2 and $\overline{E}_2$ with the state operator σ which is trivial since the normalization of the state operator is irrelevant for additivity. The only other couple is $a = (E_1, \overline{E}_2)$, $b = (\overline{E}_1, \overline{E}_2)$, which is analogous to the previous one and gives the condition $\mathrm{Re}\,\mathrm{Tr}(E_1 \rho \overline{E}_1 \overline{E}_2) = 0$. This reduces, however, to equation (14.1) because of the relation $E_2 + \overline{E}_2 = I$, which gives $\mathrm{Tr}(E_1 \rho \overline{E}_1 \overline{E}_2) = \mathrm{Tr}(E_1 \rho \overline{E}_1) - \mathrm{Tr}(E_1 \rho \overline{E}_1 E_2)$. But from cyclic invariance of the trace one has $\mathrm{Tr}(E_1 \rho \overline{E}_1) = \mathrm{Tr}(\rho \overline{E}_1 E_1)$, which vanishes since $\overline{E}_1 E_1 = 0$, from which equation (14.1) follows again.

118. For any family of histories, the previous procedure gives a set of consistency conditions as necessary and sufficient criteria for additivity. These conditions can always be written similarly: as a vanishing real part of the trace of a product involving history operators and the state operator.

We already saw how a field of propositions can be associated with a family of histories, including histories, properties and various other propositions that can be obtained with and, or, and not. When the consistency conditions are satisfied, inference and logical equivalence are also well defined and all the axioms of ordinary logic hold.

The consistency conditions therefore provide a criterion for selecting the families of histories in which logic makes sense. This result is remarkable because it shows that a few algebraic equations (the consistency conditions) select which kind of discourse makes sense and which ones are nonsense.

119. The Griffiths consistency conditions are necessary and sufficient for the validity of logic. A few years of practice with the history language have shown, however, that they can be conveniently replaced by another criterion, which was proposed by Gell-Mann and Hartle. In the case of our example, there is a unique GMH condition, namely

$$\mathrm{Tr}\left[E_1(t_1)\rho\overline{E}_1(t_1)E_2(t_2)\right] = 0. \tag{14.2}$$

Comparing it with equation (14.1), one sees that not only the real part of the trace is supposed to vanish but also the imaginary part. The GMH condition is therefore stronger, and in fact equation (14.2) implies equation (14.1). One can therefore be sure that logic is valid when the GMH condition holds. It is sufficient though not necessary.

There are several reasons for preferring the GMH conditions in the practice of interpretation: (1) They are easier than the Griffiths conditions to write out, although this simplification occurs only in cases with three or more reference times. (2) The GMH conditions hold almost automatically when the histories are suggested by classical physics or when they rely on decoherence. These are precisely the essential ingredients of an understanding of interpretation, and the GMH conditions can often be justified by intuitive arguments with no involved computation. (3) Even in the case of a finite-dimensional Hilbert space, the difference between the two types of conditions seems rather irrelevant because no one has yet found an interesting problem for which equation (14.1) would yield something more instructive than equation (14.2).

Example

One can consider as an example a spin 1/2. It is prepared in a pure state, which is an eigenstate of the spin component along a direction n_o, that is, $S.n_o = 1/2$. A family of histories includes the properties of the spin component along a direction n_2 at a time t_2, which can be the results of a measurement with a Stern-Gerlach device. An interesting question is to decide what can be said about the spin at an intermediate time t_1 that is later than preparation and earlier than measurement.

One assumes the Hamiltonian H to be zero. Every projector in a two-dimensional Hilbert space can be written in terms of the Pauli matrices and a space direction n_1 as $(1/2)(I + \sigma.n_1)$. One can then say that the consistency conditions select the directions n_1 that can enter in a rational discussion of the situation at time t_1.

The calculation of the trace in equation (14.2) is straightforward. Writing $\rho = (1/2)(I + \sigma.n_o)$ and using the identities $(\sigma.n)(\sigma.n') = n.n' + i\sigma.(n \wedge n')$ and $\mathrm{Tr}\,\sigma_i = 0$, equation (14.1) gives $(n_o \wedge n_1).(n_1 \wedge n_2) = 0$, whereas equation (14.2), assuming also a vanishing imaginary part, adds the condition $n_o.(n_1 \wedge n_2) = 0$. Altogether, using condition (14.2), these two equations imply that n_1 is either parallel to n_o or to n_2.

The GMH condition reduces the discussion to one concerning complementarity, which will be dealt with in the next chapter: At time t_1, is the spin still oriented in the direction in which it was prepared, or is it already in the direction where the measurement will find it later?

The Griffiths condition requires only the geometric relation $(n_o \wedge n_1).(n_1 \wedge n_2) = 0$, which is satisfied by $n_1 = n_o$ or n_2 and also by a whole set of directions with no obvious physical meaning. Moreover, except for the two directions n_o and n_2, no direct inference from the preparation or the measurement results can afford a precise statement of some spin properties in these directions.

A GENERAL FORMULATION

120. I will now summarize how one can use the language of histories in a rational way. In the general case, one is dealing with a family of mutually exclusive histories whose representative operators C_a satisfy the completeness relation

$$\sum_a C_a = I. \tag{14.3}$$

I will not write the general form of the Griffiths conditions generalizing equation (14.1). As for the GMH conditions, they can generally be written as

$$\mathrm{Tr}(C_a \rho C_b^\dagger) = 0, \tag{14.4}$$

for any couple of mutually exclusive histories a and b. They imply the property of additivity for the probabilities of histories, but the proof is nontrivial and will not be given here.

One may notice that the last remaining axiom of probabilities, namely the normalization condition

$$\sum_a p(a) = I,$$

is also satisfied

Proof

According to equation (14.4), one has $\Sigma_a \mathrm{Tr}(C_a \rho C_a^\dagger) = \Sigma_{a,b} \mathrm{Tr}(C_a \rho C_b^\dagger)$, which can also be written as $\mathrm{Tr}((\Sigma_a C_a)\rho(\Sigma_b C_b))$ or according to equation (14.3), $\mathrm{Tr}(I\rho I)$, which is 1.

VERIFYING THE CONSISTENCY CONDITIONS

121. One might feel anxious at the prospect of having to write out explicitly the consistency conditions (14.4) and then to compute all the necessary traces in a case of physical interest. We saw an example for a spin $1/2$ in the previous section, that involved the minimum number of reference times, a two-dimensional Hilbert space, and a trivial Hamilto-

nian $H = 0$. Nothing could seem simpler, yet the calculation was not completely trivial.

It might look as if interpretation has been reduced to a well-defined algebraic theory, but at the price of involved calculations. A theorist will immediately point out a formidable feature of consistency conditions: they involve twice as many projectors as one would take into account when computing the evolution of a wave function. Does it mean that interpretation is condemned to make the difficult calculations of quantum mechanics twice as hard? The question is worth additional comment.

There are five methods for computing the trace in a consistency condition. The first one is to perform a straightforward algebraic calculation, as in the previous example. It works at best in a finite-dimensional Hilbert space with a low dimension.

The second method begins as if one were to perform a complete calculation. Each projector $E(t)$ in a history operator C_a is written explicitly as $U(t)^{-1}EU(t)$. The explicit form of the evolution operator is used, and one has, for instance, in the case of a free nonrelativistic particle,

$$\langle x|U(t)|x'\rangle = (m/2\pi i\hbar t)^{3/2}\exp\left[im(x-x')^2/2\hbar t\right].$$

A consistency trace is therefore an integral over many variables of a rapidly oscillating quantity with a well-defined and finite integration range. The oscillating character originates in the various evolution operators $U(t)$. The number of integration variables increases rapidly with the number of particles in the system and the number of reference times in the history. A property concerning position requires that an integration is performed on a bounded set. The same is true for a property concerning momentum with the supplementary burden of performing a Fourier transform. Things are more complicated when the particles interact or have a relativistic motion and consistency integrals can only be computed by means of approximate methods, particularly the saddle-point method, which relies on the smallness of Planck's constant. These cumbersome methods were used in the early days of histories (Omnès, 1994a) but they cannot be recommended.

The third method consists in replacing the evolution operators by Feynman path integrals. It may look like a clumsy trick to replace a difficult integral with many integration variables by another one involving infinitely many variables. One can, however, use specific approximation methods (Feynman and Hibbs, 1965) that also often rely on the smallness of Planck's constant.

When the smallness of Planck's constant is helpful, it means that classical physics is not far from being satisfied, and it is tempting to use this simpler physics directly. This is the principle of the fourth method. Classical considerations are used in many places when an experiment is conceived, and the construction of experimental devices is also often thought of in classical terms. Particles beams, for instance, are built on the idea of a classical trajectory. A wire chamber assumes straight-line tracks for charged particles. The same is true when systematic errors are estimated by using a Monte-Carlo method. The examples are numerous and ubiquitous, and classical physics determines the choice of many relevant properties entering the practice of physics.

The fourth method consists in using directly the theorems from Chapters 10 and 11 regarding the validity of classical physics and error estimates. The Egorov theorem shows that in a consistency trace, many classical events are tautologies if they follow from determinism, as if the corresponding projectors were the identity operator (determinism is a tautology). When two exclusive classical events enter in two histories a and b, the product $C_a C_b^\dagger$, is exponentially small and the consistency trace vanishes. The power of this method is therefore to dispose of many computations by reducing the validity of a consistency condition to the common sense of physics. More exactly, it relies on the proofs of the powerful microlocal theorems that were mentioned earlier.

The last and fifth method also relies on very general results. It is based on the decoherence effect, which implies consistency with extremely small errors in many cases of great physical interest. It will be described later and it is particularly useful in the case of measurement.

One may also mention that the choice of the reference times $(t_1, t_2, \ldots, t_n)$ occurring in a history is often dictated by considerations of classical physics and decoherence. Their exact value is therefore not of much importance as long as they remain within commonsense limits.

We have thus reached some satisfactory conclusions: Explicit computations are almost never necessary when one is sensible enough in one's choice of discourse. The corresponding histories are practically obvious, and they need not be made explicit. Of course, some theory must be used if one wants to compute explicitly the errors arising from classical considerations or from the use of decoherence, but this can be done as in Chapters 16 and 18. To conclude briefly, *despite a potentially involved mathematical apparatus, the practice of interpretation requires very few calculations, and most often none at all.*

A CASE OF NONSENSE HISTORIES

122. It may be guessed from what was just stated that an actual computation of consistency conditions will be especially needed when something logically suspicious is under way. Computation can then be used to point out any illogical processes. I will give only one example: the well-known case of an interference experiment in which it is asserted that a particle cannot simultaneously cross the two slits in a Young device or the two arms of an interferometer. I will rely in this case on the Griffiths conditions (14.1) rather than the GMH conditions (14.4). This is because the Griffiths conditions are necessary and sufficient for ensuring logical consistency, and we want to make sure that no logical loophole can remain.

The calculations are easy if one relies on the elementary theory of interferences, which is equivalent to the semiclassical WKB method for computing wave functions. One may use a family of histories with two reference times t_1 and t_2. The initial state is assumed to be a plane wave packet ψ with average wave number k. It will be convenient to think of an interferometer with arms long enough for the wave function to be split between them at time t_1. The screen on which the particles arrive (e.g., a photographic plate) is divided ideally into many regions labelled by an index j, small enough when compared with the interference pattern. The particles arrive on the screen at time t_2.

Let $\phi_j^{(1)}$ denote the phase of the wave function arriving at point j after crossing the first arm of the interferometer and $\phi_j^{(2)}$ in the case of the second arm. Let a be the history stating that the particle crosses the first arm before reaching j, and b the history for going through the second arm and also arriving in j. Let $E_1(t_1)$ and $E_1'(t_1)$ be the two projectors expressing that the particle position is respectively in one arm or the other at time t_1. The fact that the wave function $\psi(t_1)$ is split between the two arms is expressed by the mathematical relation

$$(E_1(t_1) + E_1'(t_1))\psi = \psi.$$

Rather than writing out the Griffiths consistency condition (14.1), it will be more convenient to go back to the calculation from which it was derived. Let $c = a$ or b, which is also a history with the projectors $(I, E_{2j}(t_2))$. The probabilities of the three histories a, b, c are computed in textbooks on interference theory and are given by

$$p(a) = A, \quad p(b) = A, \quad p(c) = A\left|\exp\left(i\phi_j^{(1)}\right) + \exp\left(i\phi_j^{(2)}\right)\right|^2,$$

where A is some geometrical coefficient, which would be easy to write out. A consistency condition (there are as many of them as different regions j) expresses the additivity property $p(c)=p(a)+p(b)$, which can be written as

$$\cos\left(\phi_j^{(1)}-\phi_j^{(2)}\right)/2=0. \qquad (14.5)$$

One can write finally $\phi_j^{(1)}=kL_1$, k being the wavenumber and L_1 the length of the classical path reaching the point j after crossing the first arm, with a similar expression for $\phi j^{(2)}$. It is clear that the set of consistency conditions (14.5) cannot be satisfied because the phase difference $\phi_j^{(1)}-\phi_j^{(2)}$ varies continuously with the position of j. Logic cannot apply to this description of the experiment, which is nonsense.

APPROXIMATIONS IN LOGIC

123. We must now consider a point which is still a topic of controversy about the foundations of the history language. It was implicitly assumed up to now that a valid inference relies on a conditional probability equal to 1, by which one means strictly equal to 1. Similarly, the consistency conditions ensuring the validity of logic suppose some traces to be strictly zero. It turns out, however, that the consistency conditions occurring in practical applications are satisfied with a high degree of approximation, but not exactly. Consistency traces as in equation (14.4) are found to be very small but not exactly zero. Similarly, some conditional probabilities $p(a|b)$ underlying an inference $a \Rightarrow b$ are not exactly equal to 1 but only very close to 1, say $1-\varepsilon$ with ε very small.

The reason is clear. When one is using classical physics, the consistency conditions and the implications are controlled primarily by the exclusion formulas between classical properties involving exponentially small errors, such as in equation (10.13). Even smaller errors are met when consistency and inferences come from a decoherence effect, although the errors are not yet strictly zero.

Some critics would say that the precision is excellent but not perfect. It does not agree with a tradition of exactness in formal logic, which is not respected in the present case. One might try to reach perfection by using the formal Griffiths consistency conditions rather than the more physical GMH conditions. It is then often possible to satisfy the conditions exactly because a consistency trace oscillates very rapidly with the parameters it involves, for instance, the boundary of the range in a property. By giving a very well-chosen and precise value to the parameters, one can satisfy exactly the consistency conditions (as shown in Omnès, 1994a, Chapter 5, Section 9). As for a conditional probability equal to $1-\varepsilon$, it is interpreted as underlying an inference $a \Rightarrow b$ with a probability of error ε, an idea which looks quite reasonable when the propositions a and b are expressing physical statements.

All this painstaking exertion for exactness is, however, hair-splitting. I, for one, may pay all due respect to a logician asking for exactness, but nevertheless would not dare interrupt one of my experimentalist friends during a seminar and ask: "Did you notice that the beam in your experiment is not exactly classical? A particle can enter from the front into your measurement device but it could also come from behind by a wave diffraction effect. Your conclusion on the blackboard is not therefore strictly correct!" The friend would grin and ask: "By how much do you estimate my slip in reasoning?" Were I to answer "As much as ten to the power minus 144," I would hate to hear the laughter, and the corroboration of a logician would not be of much help.

More seriously, one should not forget that the purpose of interpretation is to understand what is going on in an experiment and to express it *reliably*. "Reliably" does not mean "perfectly." It would be useless and boring to unceasingly repeat: "What I say is valid with an error such and such," or "I call your attention to the fact that my statement is only ideally valid if the boundary of the laboratory containing this particle is defined with a perfect precision." Everybody knows that although quantum fluctuations exist, they are completely negligible in many circumstances. Who wants more?

A RULE FOR INTERPRETATION

124. We have reached a point from which interpretation can become a deductive theory. It will be enough for that purpose to state the following rule:

Rule. *Every description of a physical process can be expressed by means of a unique complete family of consistent histories. Reasoning on the physics of the system can always be done by inferences that can be proved in the framework of this family.*

When one is particularly interested in a specific history or a specific process, one can always formally build a complete family around it by introducing the negation of every property in the history. This family must be consistent for the history to make sense. It was proposed to elevate this rule to the level of a principle of interpretation that is on a footing similar to the basic rules of quantum mechanics (Omnès, 1992, 1994a). It seems better, however, to take it as a guiding rule in the maze of interpretation, in keeping with our understanding of histories as a constituent of language.

The reason the rule insists on a unique family is the result of complementarity, which will be explained in the next chapter. The kind of physical reasoning that is mentioned in the rule can be explained with an example. Recalling the nuclear reaction $n + p \rightarrow d + \gamma$, we can overhear the discourse of a physicist mumbling to himself (or herself): "The photomultiplier has clicked. Therefore a photon entered it. Therefore a reaction producing the photon has occurred. It was produced by a neutron, which went necessarily through the velocity selector. Therefore I know the initial neutron momentum. The proton momentum is very low according to the preparation of the target, and the direction of the photomultiplier can be used to obtain the photon momentum. Therefore I know the kinematics of the reaction which occurred and this is what I wanted."

The everyday life of physics is replete with arguments of that kind, often trivial but sometimes very important. They are essential, however, they look far from obvious if one thinks of them in the light of the basic principles. A great virtue of the history language is therefore to justify, once and for all, what looked obvious and necessary from the standpoint of practice and was never before fully analyzed from the standpoint of theory.

SUMMARY

The probabilities relative to a complete family of histories must satisfy some consistency conditions to be genuine probabilities. The most convenient sufficient conditions are given in terms of the history operators by $\mathrm{Tr}(C_a \rho C_b^\dagger) = 0$ for any couple of different histories (a, b). Although these conditions look complicated when explicitly written out, they can be justified with practically no calculation in most cases with the help of classical considerations and decoherence effects.

The best rule for soundly using the language of interpretation consists in describing the possible events as a physical process within the framework of a unique complete family of consistent histories. The rational arguments necessary for the understanding and exploiting of an experiment or a natural process relies on probabilistic inferences ($p(b|a) = 1$ standing for a logical inference $a \Rightarrow b$). With these conventions, interpretation can be turned into a deductive theory.

15

Complementarity

125. The rule for interpretation I gave in the previous chapter provides the language of interpretation with a definite logical framework. The language, which we have seen includes the standard language of classical physics, has however, a very peculiar feature when it applies to a microscopic object such as an atom or a particle. This feature is complementarity.

Bohr gave a central role to complementarity in the language of interpretation, and as a matter of fact, he practically replaced everything in the language of physics with this unique feature. This looked, however, like a rather negative idea, as when he said: "Any given application of classical concepts precludes the simultaneous use of other classical concepts which in a different connection are equally necessary for the elucidation of phenomena" (Bohr, 1934, p. 10). More trivially, but perhaps more emphatically, we already mentioned another version: "When I hear this guy on the phone, he speaks like a man, but when I see him, he looks like a cat and does not say a word."

Complementarity is clearer in the framework of histories. It is not a new principle but rather a consequence of the basic ones, amounting to recognizing that *several different consistent families of histories can describe the same physical process without being mutually consistent.*

Complementarity has always been found to be an obstacle to attributing reality to the quantum world. Although this is a deep philosophical question on which much has been written, I will only discuss it briefly. One must avoid the common confusion between reality itself and the way we think of it. To say that the experimental setup I am looking at is "real" means in order of increasing precision: (1) The setup exists by itself, independently of my presence, of my looking at it, and my thinking about it. (2) I can perceive and see the phenomena it involves. We often call these phenomena "facts" and consider them as the building blocks of any empirical science.

From this, one must distinguish the way we think of reality. I may have dreamed of a dragon and tell my dream to my wife. In my mind, however, there is a strong difference between the dragon and my wife: only she is real. Even when I do not perceive her presence, I consider her as real. There is in my mind, as Kant would say, a category of reason that is the category of reality, the idea of something possessing reality. To this idea I can apply the two previous characteristics (1) and (2). Often there is a third one, which is (3) I might describe every part of this system as precisely as I wish and under every aspect I wish, at least in principle and in a logically consistent way.

This third feature is the critical one, which gave rise to so many controversies. It is true as long as one remains at a macroscopic level (excepting tricky macroscopic nonclassical systems built in a laboratory). It does not apply at the quantum level, as shown by complementarity, and we must get accustomed to that exception.

I will give several examples of complementarity, including a famous one by Einstein, Podolsky, and Rosen. At this point, however, I feel embarrassed to decide whether these questions are important for every physicist or not and if readers should be advised to read this chapter or skip it. It depends on one's own inclination, but one can say at least that complementarity has no relevance for the *practice* of physics. It shows an interesting feature of the *language* of interpretation, which is unusual in other sciences.

A DEFINITION OF COMPLEMENTARITY

126. Two families of histories (or two logical frameworks) will be called complementary when they are both consistent though mutually incompatible. The existence of such exclusive frameworks is obvious in view of many examples. It was anticipated in the rule of interpretation I gave in Section 124 when I insisted on using a unique consistent family of histories in order to avoid apparent paradoxes.

The origin of complementarity lies in the noncommutativity of operators and particularly of projectors. At a given time, for instance, one cannot speak of both position and of momentum, except when the error bounds are so large that classical physics applies. When classical physics does not apply, one must accept complementarity, which is certainly the most specific character of quantum logic.

One can say more precisely that two families of histories F and F' are complementary when they are both consistent and there is no consistent family including both of them, i.e., containing all the properties occurring either in F or F'.

A simple analogy of this feature of quantum logic with everyday logic appears in the domain of the law. There is little common content, for instance, between the legal rules for property in our modern civilization and the customary law of some older traditional cultures. Bohr also drew some analogies with the difference between a vitalistic and a functional account of physiology, or the approach to consciousness through introspection or brain physiology. Bohr gave several such examples, but their multiplicity could draw a naive reader of Bohr toward a type of relativism in philosophy, a skepticism that could be easily extended to the laws of nature themselves, whereas in the present approach, the *knowledge* of these laws is responsible for a dire limitation of natural languages. I believe, therefore, that a proper attitude is to consider complementarity as a specific character of quantum mechanics without extensions to other domains.

A No-Contradiction Theorem

127. It goes without saying that if complementarity could entail logical contradictions, it would be an insuperable hindrance for interpretation. This danger is fortunately removed by a no-contradiction theorem, which is as follows:

Theorem. *Let a and b be two propositions belonging to a consistent family of histories in which the inference* $a \Rightarrow b$ *is valid. If another consistent family also contains these two propositions, the same inference is valid in its framework, whether or not the two families are complementary.*

The theorem looks rather obvious since it deals with the two propositions themselves. Its proof, however, is not quite trivial because of the constructions one needs for relating the geometrical representations of each family by some sets in a specific many-dimensional space. A rather provocative formulation of this theorem would be to say that no paradox can ever happen in quantum logic. As a matter of fact, this is a true statement, but complementarity results in some peculiarities. A few striking examples will now be given.

EXAMPLES

An Example with a Spin 1/2

128. Let us recall the example in Section 119: A spin 1/2 is prepared at time zero in a pure state with $S_x = 1/2$. One measures the spin component S_z at time t_2, which leads, for logical purposes, to the introduction of the two properties $S_z(t_2) = \pm 1/2$. What can be the properties of the spin at an intermediate time t_1? We saw that the GMH

criterion (14.4) suggests two different types of properties at time t_1, involving either the pair of properties $S_x(t_1) = \pm 1/2$ or the pair $S_z(t_1) = \pm 1/2$. The two families one obtains are consistent but complementary. There is no reason to prefer one or the other, and they are never found to be necessary in any relevant discussion of the underlying physics.

Straight-Line Motion

In this example, a heavy radioactive nucleus is located at the origin of space O and it produces at time 0 an alpha particle with mass m in a S-wave (i.e., with zero angular momentum with respect to the origin). The emitted particle is seen at time t_2 by a detector occupying a small region of space V_2 centered at a point x_2 (Figure 15.1). Since the beginning of quantum mechanics, the following question has been asked: does the alpha particle at an intermediate time t_1 still have zero angular momentum, or is it already on its way toward x_2 along a straight-line trajectory?

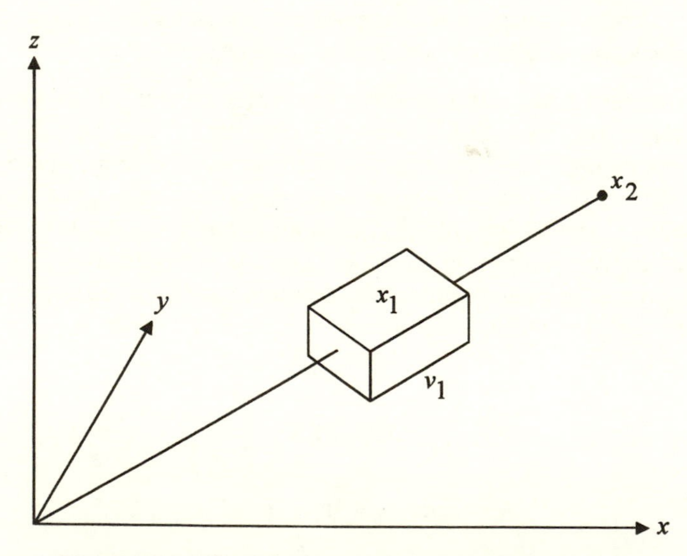

Figure 15.1. Straight-line motion. A particle emitted isotropically at the origin is detected at the position x_2. The question is: was the particle going toward x_2 and located in a region V_1 near a point x_1 at an earlier time?

To make the discussion more precise, one can assume that the initial radial velocity v of the alpha particle is rather well defined by the decay kinematics. Knowing that the particle reaches x_2 at time t_2, one can compute the point of space x_1 where it should have been at time t_1 according to classical physics. Let V_1 be a sufficiently large region of space centered at x_1. The idea of a straight-line motion can be expressed by the statement: "the alpha particle position is in V_1 at time t_1." Let us therefore introduce this property as well as its negation into a family of histories, together with the property "the particle position is in V_2 at time t_2" and its negation. One thus gets a two-time, yes-no family, with which we are already familiar.

Is it consistent? The correspondence theorems in Chapters 10 and 11 give a positive answer. One can also obtain the following inference: if the particle position is in V_2 at time t_2, then it is in V_1 at time t_1.

If this is not obvious, one may introduce a classically meaningful projector E at time zero, ascertaining the position in a neighborhood of the origin and the momentum in a spherical shell with a medium radius mv. Assuming an initial wave function ψ, it satisfies $E\psi = \psi$. It is classically obvious that if the particle does not go through V_1, it will never reach V_2 (when one says "obvious," one is of course relying ultimately on Egorov's theorem). This means that the product $E_2(t_2)\bar{E}_1(t_1)E$ is zero, up to negligible corrections, and the validity of the GMH consistency condition follows immediately. It is also clear that

$$E_2(t_2)E_1(t_1)\psi = E_2(t_2)E_1(t_1)E\psi = E_1(t_1)E\psi = E_1(t_1)\psi$$

from which one obtains the conditional probability $p(E_1(t_1)|E_2(t_2)) = 1$, showing that $E_2(t_2) \Rightarrow E_1(t_1)$.

Let us now consider the second family, assuming the alpha particle to have still angular momentum zero at time t_1. It is also perfectly consistent, but the two families are complementary.

Let F be the projection operator on angular momentum 0. One has $F\psi = \psi$, and furthermore, F commutes with the Hamiltonian since angular momentum is a constant of motion. One has therefore $\bar{F}_1(t_1)|\psi\rangle\langle\psi|F_1(t_1) = 0$, from which the GMH consistency condition follows. The probability for the angular momentum to be 0 at time t_1 is immediately seen to be 1.

The question of straight-line motion was considered for the first time by Mott (1929a), who asked why a charged particle can leave a straight-line track in a Wilson ionization chamber in spite of its description by a wave function. Mott's calculation might be used directly for deriving the results we have obtained just now, and they would provide an example of the second method of computing consistency traces and probabilities as indicated in Section 121.

There is another way of stating the existence of a straight-line motion. It deals with velocity rather than position. The intermediate property one introduced states that the velocity is directed toward x_2 at time t_1 within some reasonable error bounds. The family of histories one obtains is again consistent, and the velocity property is again a logical consequence of the observation. Of course, there is complementarity with the previous family stating a property of position at the same time.

THE QUESTION OF REALISM

129. The last example shows that one cannot consider as strictly real the straight-line motion of a particle when there is no direct observation (measurement) of this motion. Similarly, one cannot assert that the particle still has its initial angular momentum or that, rather than a grossly localized particle, the system is "really" an expanding wave. As a matter of fact, these various points of view are sensible, but complementarity shows that they are incompatible.

There should be no doubt, of course, that something exists inside the experimental setup, that does not depend on our own minds. In other words, that "something" we may sometimes call a particle and sometimes call a wave is real. The trouble is that we do not know what the category of reality means in that case because it has lost its uniqueness even though the "real object" exists.

This concept is not completely new and philosophers realized that reality cannot be defined. According to Ludwig Wittgenstein, something is real when one can point a finger at it and say "that." This is of course the ordinary reality we see everywhere in our macroscopic surroundings, and it is obvious as a pure experience. Our language can describe it and help us think about it, but it cannot define it except for classifying the corresponding ideas in the category of what is real.

The loss of a unique and complete description of physical reality by the language of physics has been certainly the worst blow to the representation of the world that has arisen from quantum mechanics. It originates in noncommutativity, just like the Heisenberg uncertainty relations implying the loss of a visual representation. Einstein could never consent to such a sacrifice. As for Bohr, he said: "We must never forget that 'reality' too is a human word just like 'wave' or 'consciousness.' Our task is to learn to use these words correctly" (quoted by

Kalckar in Rosental, 1967, p. 227). Elsewhere, he wrote: "In our description of nature the purpose is not to disclose the real essence of the phenomena but only to track down, so far as it is possible, relations between the manifold aspects of our experience" (Bohr, 1934, p. 18).

A SPURIOUS EXAMPLE

130. Complementarity is not always easy to deal with. It can lead to strange consequences when one forgets that consistent histories should provide a relevant language and one instead makes them an occasion for gratuitous mathematics. An example proposed by Adrian Kent (1997), who elaborated on a remark by Yakir Aharonov and Lev Vaidman (1991) having no relation to histories, will show how tricky or subtle these questions can be.

One considers a spin 1 system. It is prepared in a pure state $|0\rangle$ having the quantum number $m_o = 0$ for the spin component along a space direction n_o, that is, $S.n_o = 0$. At time t_2, one measures the spin component in a direction n_2. There is also an intermediate time t_1 and the properties to be considered at that time are somewhat unusual since they depend on the result of the measurement. If this result is $S.n_2 = 0$ (corresponding to a vector $|2\rangle$ in Hilbert space), one considers a direction n_1 and the property "$S.n_1 = 0$" (associated with a Hilbert space vector $|1\rangle$) as well as its negation at time t_1. If the result of the measurement is $S.n_2 = \pm 1$, no statement at all is to be introduced at time t_1 (i.e., one has only the property with projector I).

One can construct a corresponding family of histories. Introducing the projectors $E_1 = |1\rangle\langle 1|$, $E_2 = |2\rangle\langle 2|$, the family is composed of the three histories $(E_1, E_2), (\bar{E}_1, E_2), (I, \bar{E}_2)$. The Hamiltonian is supposed to be $H = 0$. There is only one Griffiths consistency condition, which can be considered as an algebraic equation for the direction n_1, and this equation has sometimes two orthogonal directions n' and n'' as solutions. When one chooses either n and n' for the direction of n_1, one gets two complementary consistent families of histories.

Then comes an apparently curious result. The inference "$S.n_2(t_2) = 0$" $\Rightarrow$ "$S.n(t_1) = 0$" is valid in the first family. But in the second family, one has similarly "$S.n_2(t_2) = 0$" $\Rightarrow$ "$S.n'(t_1) = 0$." It happens, however, that the two Hilbert space vectors $|S.n = 0\rangle$ and $|S.n' = 0\rangle$ are orthogonal. One thus obtains two orthogonal properties (projectors) as a consequence of the same measurement in two complementary frameworks.

*The calculation**

Let us first establish a correspondence between a space direction n and the Hilbert space vector representing the property $S.n = 0$ in the case of a spin 1. The Hilbert space has 3 (complex) dimensions. It provides an irreducible representation of the rotation group in three-dimensional Euclidean space where an infinitesimal rotation around the z-axis in space is represented by the matrix iS_z. There is a basis such that this matrix transforms a vector with complex components (x_1, x_2, x_3) in Hilbert space into $x'_1 = -x_2$, $x'_2 = x_1$, $x'_3 = 0$. The eigenvector associated with $S_z = 0$ is therefore $(0, 0, 1)$. More generally, an eigenvector corresponding to the eigenvalue equation $S.n = 0$ has the same components in Hilbert space than the vector n in space and the corresponding projector is represented by the same matrix as the tensor product $n \otimes n$. When two space vectors n and n' are orthogonal, the Hilbert space vectors associated with $S.n = 0$ and $S.n' = 0$ are orthogonal.

The consistency condition (14.1) can then be written out, noticing that the initial state operator is represented by the matrix $n \otimes n$. It is given by

$$\mathrm{Re}\,\mathrm{Tr}(n_1 \otimes n_1, n_o \otimes n_o.(I - n_1 \otimes n_1).n_2 \otimes n_2)$$
$$\equiv (n_1.n_2)(n_1.n_o)(n_o.n_2) - (n_1.n_2)^2(n_1.n_o)^2 = 0,$$

which gives, after an explicit algebraic calculation, $n_o.n_2 - (n_1.n_2)(n_1.n_o) = 0$.

Let one take $n_o = (1, 0, 0)$, $n_2 = (\cos\theta, \sin\theta, 0)$, $n_1 = n_\pm$, with $n = n_+$, $n' = n_-$ and $n_\pm = c(\cos\theta, a^{-1}\sin\theta, \pm a^{-1}\sqrt{a-1}\sin\theta)$, a being a real number larger than 1 and c a normalization coefficient. The previous condition is then automatically satisfied. For the two vectors $n_\pm$ to be orthogonal, one must have $a^2\cos^2\theta + (2-a)\sin^2\theta = 0$, an equation in a which has real roots if $\cos\theta < 1/3$.

Is this result self-contradictory? Certainly not, because nowhere did one consider two properties in two complementary families as conflicting. One might object that this answer is somewhat formal. But, histories are meant to provide a language and a language can certainly express irrelevant (although not paradoxical) ideas. There is something irrelevant in this example, by which one means that it would never occur in a sensible description of an experiment.

As a matter of fact, the most general structure, of relevant histories is the GMH structure where a property at a time t_k can only be an occasion for choosing some relevant properties at the next time t_{k+1}. The present families are not of that type: The property $S.n_2 = 0$ at time t_2 is taken as an occasion for introducing the property $S.n_1 = 0$ (whatever the direction n or n' of n_1) at a *previous* time t_1, while nothing of that sort is considered at time t_1 if $S.n_2 = \pm 1$. Logical branching is taken in the wrong direction.

131. One can be less formal by using a method due to Heisenberg in the framework of the Copenhagen interpretation (see, e.g., Heisenberg, 1930). It consists in giving substance to a microscopic property by envisioning an imaginary measuring device as being present in the right place and active at the right time for checking it.

It is assumed that three measuring devices M, M', and M_2 are present in the experimental setup, the first two being imaginary and the third one real. The real one, M_2, measures the spin component along n_2 at time t_2. It is, for instance, a Stern-Gerlach device. Each imaginary device M or M' can be eventually activated at time t_1. M is not exactly a Stern-Gerlach device measuring the spin component along the direction n, but rather a yes-no device testing for the validity of the property $S.n = 0$ at time t_1. It could be realized in principle by means of a Stern-Gerlach magnet and a detector (which can be activated or not) standing in the way of the trajectory of the spin state $S.n = 0$. The detector checks the passage of an atom without disturbing it. The three different trajectories for the states $S.n = 0, \pm 1$ behind the magnet are brought back together by means of some clever magnetic device, and they join in a unique trajectory before entering the second device M'. It is identical to M except for the direction of the magnet, which is now n'. After recombination of the trajectories, the atom finally enters the device M_2, which is a real Stern-Gerlach device (see Figure 15.2).

If Kent's example were a real stumbling block for the language of histories, it would really be overkill because it would also destroy the Copenhagen interpretation, at least Heisenberg's version. On the contrary, one can see with Heisenberg's method what is wrong in the upside-down order of logical branching: What does it mean to activate M (or M') at time t_1 when and only when the measurement result $S.n._2 = 0$ is obtained at time t_2? The direction of time is inverted and the experiment is impossible.

A critic could still argue and say that in the examples given in Section 128 the localization property in the region V_1 was also introduced after the detection at point x_2. This was essentially the same example as Kent's, except for the orthogonality of the projectors. Kent's example succeeds in turning the standard example of complementarity into another in which the two complementary states are exactly orthogonal, but in the end it is the familiar story that has been known for so long. Once again, the motto for avoiding the pitfalls of complementarity should be the relevance of the properties entering into a useful and instructive description of an experiment. The language of interpretation

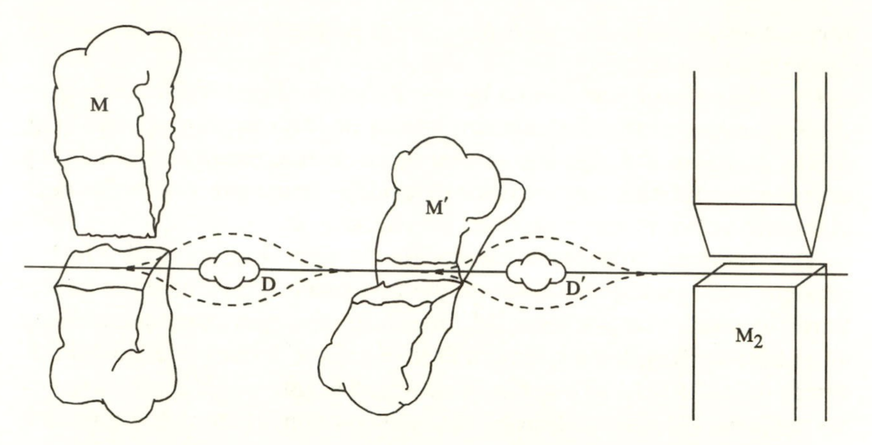

Figure 15.2. The two imaginary measuring devices M and M' (with their imaginary detectors D and D') and the real measuring device M_2 as mentioned in the text.

is not meant for expressing fantasies, although it can do it like any language.

HISTORIES AS IDEAL MEASUREMENTS

132. The previous discussion suggests a generalization of the Heisenberg method that can be used to look at histories differently. It will help narrow the gap between the two approaches and approach the Copenhagen interpretation more closely.

Let us consider for this purpose a quantum system S and a complete family of histories (which may be relevant or not for interpretation). I will denote by $\{E_j(t_k)\}$ the set of all the projectors occurring in the histories of the family, considering for simplicity a Griffiths family rather than a general GMH family.

Let us now suppose that various *imaginary* measurement devices M_{jk}, as devised by von Neumann and described in Section 49, are added to the system. The device M_{jk} measures the observable $E_j(t_k)$ at time t_k and its "pointer" indicates a position 1 (respectively 0) if it finds the eigenvalue 1 (respectively 0) of $E_j(t_k)$. At the last reference time t_n in the histories, the devices can be realistic, macroscopic ones. One may then consider the results that have been obtained from all the measuring devices, imaginary as well as real. All those having their pointer in position 1 provide in principle a signature for a definite history. The relation of this Copenhagien approach to histories with the consistency conditions is given in the Comments at the end of this book.

One sees how close the language of histories and the Copenhagen interpretation can be in practice, at least in Heisenberg's version. This is why one may say that the history approach, although it was initially independent of the Copenhagen approach, is in some sense a more elaborate version of it. It has, of course, the advantage of being more precise, of including classical physics, and of providing an explicit logical framework for indisputable proofs. But, when the Copenhagen interpretation is completed by the modern results about correspondence and decoherence, it essentially amounts to the same physics.

THE EINSTEIN-PODOLSKY-ROSEN (EPR) EXPERIMENT

133. The last example of complementarity is concerned with an experiment that was first conceived by Einstein, Podolsky, and Rosen (1935) and will be considered here in a later version by Bohm (1951). One considers two spin-1/2 particles P and P' (Figure 15.3). A measuring device M can measure the spin component of P along a direction n, and another device M' can measure the spin component of P' along n'. The two instruments can be arbitrarily distant. The particles are prepared in a pure state with total spin zero, for instance, through the decay of an unstable particle Q. One knows that whatever the direction n, the state vector for a total spin zero can be written as

$$\frac{1}{\sqrt{2}}\{|S.n=1/2\rangle \otimes |S'.n=-1/2\rangle - |S.n=-1/2\rangle \otimes |S'.n=1/2\rangle\}. \tag{15.1}$$

I will not enter into a detailed discussion of the apparatus, except for commenting that the initial presence of Q and the orientation of M and M' must be taken into account. If no common decision has been taken for choosing their orientations, they are generally along two different directions n and n', and the measurements are made at two different times t and t' (one can always arrange the two times to be different by choosing a convenient relativistic frame).

Let us assume that the measurement made by M precedes the one by M' $(t<t')$ and gives the result $S.n=s$ $(=\pm 1/2)$, the second one giving $S'.n'=s'$. Two complementary properties of the spin of P' can then be introduced (more or less analogous to the example of the alpha particle in Section 128). They refer to a time t'' when P' is not yet measured while P already is $(t<t''<t')$. One property asserts the spin of P' to be

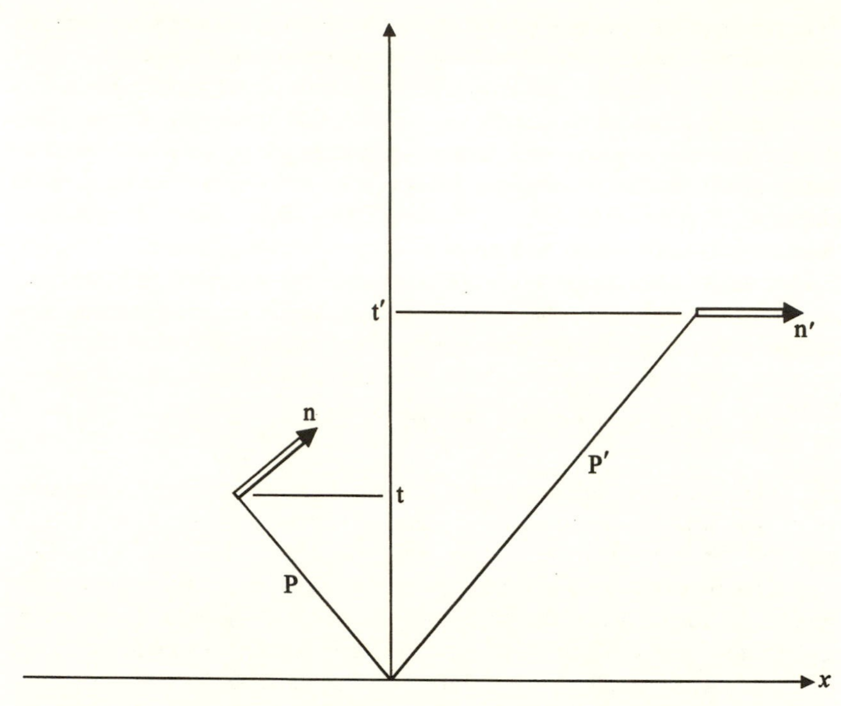

Figure 15.3. An EPR situation. The spin-component along directions n and n' of two particles P and P' are measured at times t and t'.

defined by the initial state (15.1), that is, $S'.n = -s$ (this is analogous to the property that asserted the angular momentum of the alpha particle to be $l = 0$). On the contrary, the other property anticipates the result of the measurement by M' to be $S'.n' = s'$ (this is analogous to the property assuming that the alpha particle was already going toward its future detected position along a straight line). One can then introduce the negations of the various properties to obtain two complete and complementary families of histories. They are easily shown to be consistent, and the implications $(S_P.n(t) = s) \Rightarrow (S_{P'}.n(t'') = -s)$ and $(S_{P'}.n'(t') = s') \Rightarrow (S_{P'}.n'(t'') = s')$ hold, respectively, in the two frameworks. Both intermediate assertions are therefore logically consistent although they are, of course, complementary.

One could stop the discussion there and say that neither of the two properties is really relevant for a sound understanding of the experiment, as we did previously in similar circumstances. The importance of the present experiment in the literature, however, asks for more attention, and I shall take it as an occasion for digging a bit deeper into the question of complementarity.

Einstein, Podolsky, and Rosen (EPR) were mainly concerned with the question of reality. They went as far as proposing a definition for it, or rather for an "element of reality" whose knowledge would be direct information about what really "is": "If, without in any way disturbing a system, we can predict with certainty (i.e., with probability equal to unity) the value of a physical reality quantity, then there exists an element of physical reality corresponding to this physical quantity" (Einstein, Podolsky, and Rosen, p. 777).

One may stop a moment at this point for noticing how bold this step is: *they are giving a definition of reality*! This is not a definition of the category of reality in our consciousness but rather of reality itself. No wonder, when a definition unheard of in the modern philosophy of science is opposed to the most difficult theory in physics, it could make an endless flow of papers, to which I shall try adding only a minimal amount.

After their definition, EPR proceed to show (in our example) that the property $S_{P'}(t).n = -s$ is an element of reality. It indeed gives the value of a physical quantity $S.n$ for the particle P', although only the distant particle P is disturbed by the measurement (of its own spin component along n). Looking at the state vector (15.1) and using Born's formula together with wave function reduction, one finds that the probability of $S_{P'}(t).n = -s$ is 1. There is no doubt that this is an element of reality according to EPR's definition.

EPR contended that this result implies an incompleteness of quantum theory. They said that: "the following requirement for a complete theory seems to be a necessary one: *every element of the physical reality must have a counterpart in the physical theory*" (p. 777). At this point, I will not follow the traditional discussion of the EPR experiment (intending as a conclusion the incompleteness of quantum theory) and rather envision it from the standpoint of the history language.

Let us first call attention to the fact that recourse to the reduction of the wave function, either explicit or implicit, might very well be misleading. It privileges the "element of reality" $S_{P'}.n = -1$ and a logic relying on the first measurement. It leads one to elect as meaningful the first family of histories only. But why should one prefer the indirect statement $S_{P'}(t).n = -s$ to the more direct one following from the measurement of the spin of P itself. Why not argue as follows: The spin Hamiltonian is zero, and therefore the physical quantity $S_{P'}.n'$ does not change between times t and t'. If one measures it at time t', one learns automatically what it was at time t. This is, of course, what one obtains from the second family of histories.

The most interesting case occurs when the two measurement events have a spacelike separation, in which case one can make any of them the prior one through a convenient choice of frame. The observed fact and the element of reality are permuted. Which is which?

My conclusion is that reality can certainly not be defined, nor can elements in it, because language has no power over reality and can only acceed to it through phenomena. We will return to the EPR experiment in Chapter 22 when discussing distant entangled states.

16

Recovering Common Sense

134. This chapter is devoted to several questions for which the word "essential" is certainly right. It will show how common sense, or the commonplace vision of the world does not conflict with the quantum laws of physics and can even rely on them. In this classical vision, the ordinary notion of reality is perfectly valid. When trying to understand better how this agreement is reached, we shall also see that it can open new vistas on the philosophical aspects of science.

The story began with the correspondence principle. The first proof of an agreement between classical and quantum physics goes back to Ehrenfest's theorem. This was far from sufficient, however, for a real understanding of the relation between the two physics because of two stumbling blocks: determinism and common sense.

We have seen the necessity of determinism when an apparatus works as expected or a record is deemed a fair witness of the past. Its opposition with quantum probabilism has sometimes been considered as insuperable. Common sense, on the other hand, has many facets. It involves both a global vision of the world and a specific kind of logic, as Bohr mentioned more than once. A classical vision is much more deeply rooted in our minds than Newton's dynamics. It is a totality borne by intuition (i.e., mainly a visual representation), by the language of common sense (with its primary logic), and by the notion of truth (coming from facts). Reality, as a category of thinking, is certainly what is closest to it.

A remarkable result of the modern interpretation is that everything in this sketchy description agrees with the quantum laws. This agreement will be shown in the present chapter as far as determinism and commonsense logic are concerned. Classical vision (the world we see in ordinary space) will only be recovered later as a consequence of decoherence. It should also be mentioned that friction and dissipation effects are here neglected, once again because of their strong link with decoherence.

Determinism relies on the theorems in Chapter 10 and on two simple ideas. The first one has to do with the significance of determinism, which means in a logical framework the (logical) equivalence between two classical properties holding at different times. The second idea, which cannot pretend to be new, is that determinism has a simple probabilistic expression. Rather than considering it as an absolute rule as Laplace did, one can consider it as a relation between two events whose probability of error is extremely small. The difference between the extremes of celestial physics and atoms is then found in the magnitude of the errors in determinism, extending from practically zero to approximately 1.

The logic of common sense can also be approached from a quantum point of view. It is the simplest case of the kind of quantum logic holding in the universal language of interpretation. Simplest, I say, because it deals with obvious propositions: classically meaningful properties of macroscopic objects. It is, furthermore, not hindered by complementarity so that it agrees with what we see of ordinary reality. It is really common sense.

DETERMINISM

Stating the Problem

135. Classical determinism applies to a quantum system under two conditions: (1) when the Egorov correspondence between classical and quantum dynamics is valid (see Section 97), and (2) when the initial state of the system can be described classically.

Let us make the second condition more precise. One considers a macroscopic system for which a set of collective observables has been selected. Let ρ be its initial state operator, which describes both the collective system and the (internal and external) environment. It will be convenient to use a collective (or reduced) state operator in which the environment e is ignored. It is given by a partial trace:

$$\rho_c = \mathrm{Tr}_e\, \rho. \tag{16.1}$$

This is still a state operator, positive and with a unit trace (the trace being understood in the sub-Hilbert space where the collective observables act). The two operators ρ and ρ_c give the same probability for any collective observable. One assumes the existence of a complete set of commuting collective observables that are taken as coordinates.

Consider then a classical property associated with a regular cell C_o in phase space, for instance an n-dimensional rectangle, and the property stating that "the coordinates and momenta (x, p) are given up to some maximal errors $(\Delta x, \Delta p)$." Let E_o be one of the projectors associated with C_o. The property is satisfied by the initial state at time t_o if one has

$$E_o \rho_c = \rho_c, \tag{16.2}$$

from which one obtains $\rho_c = \rho_c E_o = E_o \rho_c E_o$, using the adjoint of equation (16.2) and iterating. This is what is meant when one says that the initial state is classical. One might be more precise by giving an error for equation (16.2) or by considering the much smaller errors on the exclusion of other classical properties, but these epsilontics will be omitted.

Determinism involves the cell C deriving from C_o through classical motion during the time interval $t - t_o$. In classical physics, each of the two properties symbolized by C_o at time t_o and by C at time t imply the other one, that is, they are logically equivalent.

Let us get a bit nearer to quantum mechanism by giving a probabilistic version of determinism, although this is still classical probabilism. Suppose that the property (C_o, t_o) can occur with a probability $p(C_o, t_o)$. The deterministic generation of C when starting from C_o implies

$$p(C,t) = p(C_o, t_o) = p((C,t),(C_o,t_o)),$$

the last quantity being the joint probability for the two properties to hold. Using a conditional probability, this gives

$$p((C,t)|(C_o,t_o)) = p((C_o,t_o)|(C,t)) = 1. \tag{16.3}$$

The meaning of equation (16.3) is obvious: given that one property holds, the other one also holds with probability 1. This formulation of determinism is practically the same as that given by Laplace. It is also very close to the methods of quantum logic, and I will now exploit this analogy for finding the place of determinism in the quantum framework.

A Quantum Expression of Determinism

136. The quantum proof of determinism relies essentially on the Egorov theorem in Chapter 11. The collective Hamiltonian H_c has a Weyl symbol $h(x,p)$, which is the Hamilton function generating classical motion. The cell C is the transform of C_o through this classical motion between times t_o and t. If the conditions of Egorov's theorem are satisfied, a projector E associated with the cell C is related to E_o by quantum dynamics according to the equation

$$E = U(t-t_o)E_oU(t-t_o)^{-1} + \delta E, \tag{16.4}$$

where $U(t)$ is the evolution operator and the operator δE is small.

*Improving Determinism**

When applying Egorov's theorem to logic, one must be aware of the risks of confusion that were mentioned in Chapter 10. A projector such as E_o does not completely exclude the outside of the cell C_o and especially the points in phase space that are very close to the boundary of C_o. For obtaining the best version of determinism, it is convenient to use exclusion and introduce larger cells as in Sections 92 and 95 (equation (11.4)). One can use, for instance, a cell C' containing C as in Figure 16.1, their two boundaries being close but sufficiently separated so that the confusion of C with the outside of C' is exponentially small. The cell C' can be, for instance, an error rectangle, and it is clear that the validity of the inference $(C_o, t_o) \Rightarrow (C', t)$ will be much better than for $(C_o, t_o) \Rightarrow (C, t)$. This inference is proceeding in a definite direction of time (it is a prediction), and it represents a multitude of experimental checks of determinism (with the motion of satellites for instance).

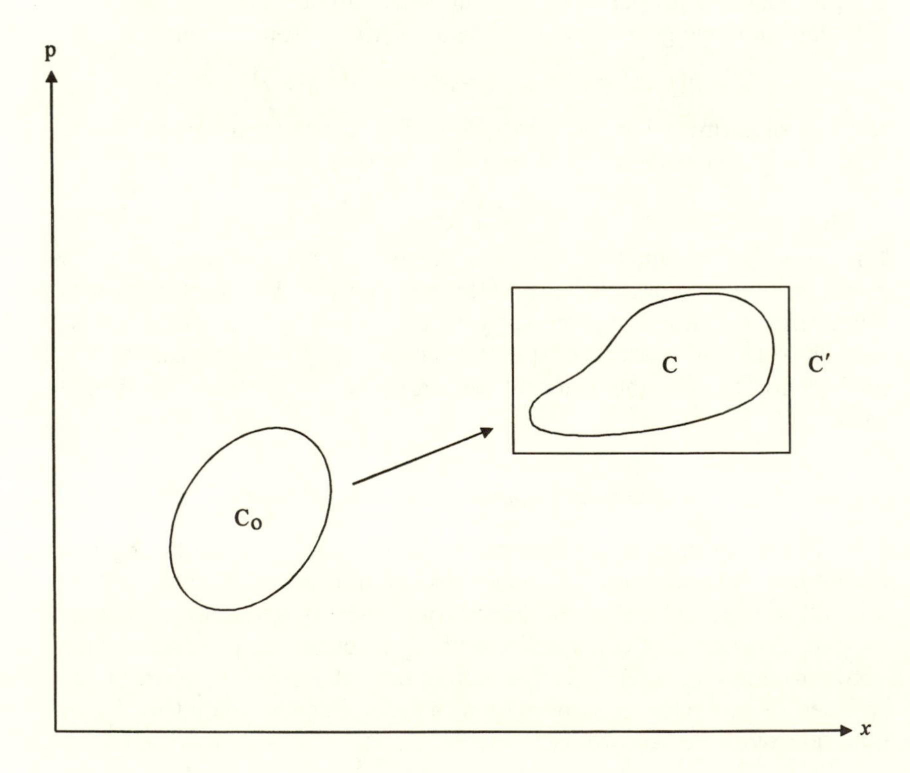

Figure 16.1. Gaining precision in determinism. The probability of error in a deterministic inference $(C_o, t_o) \Rightarrow (C, t)$ can be considerably reduced if the natural cell C is replaced by a rectangle C' enclosing it.

The gain in precision can be large when this refinement is used. If E' is a projector associated with C', equation (16.4) is replaced by another one expressing that the property (C_o, t_o) becomes part of the property (C', t). The corresponding relation is

$$U(t-t_o)E_oU(t-t_o)^{-1}.E' = U(t)E_oU(t)^{-1} + \delta' E. \qquad (16.5)$$

There is a similar relation for $E'U(t-t_o)E_oU(t-t_o)^{-1}$, and equation (16.5) is always valid when Egorov's theorem applies, although with a correction whose Hilbert norm $\|\delta' E\|$ is very small, much smaller than δE in equation (16.4). This norm is most often exponentially small as in the example given in equation (11.4).

The Proof of Determinism

The proof of the inference $(C_o, t_o) \Rightarrow (C, t)$ must use a complete family of histories including the relevant properties. One must check the consistency conditions and also that the inference relies on a conditional probability practically equal to 1. This kind of proof can be done once and for all, the difference between various applications appearing only in the estimate of errors. Even that is not a serious limitation because the errors are most often completely negligible (except for cases to be mentioned below). In cases when an estimate of errors is needed (for instance in the physics of mesoscopic objects), one can often rely directly on a calculation of quantum fluctuations rather than on Egorov's theorem.

The conclusion is that one can confidently rely on classical determinism. It has an extremely small probability of error in most cases of practical interest, and one can also have a control in questionable cases.

*Indications on the Proof**

One can rely on a family of dichotomic (yes-no) histories with two reference times where the projectors $(E_o(t_o), E'(t))$ represent the properties $(C_o(t_o), C'(t))$, and one also introduces the complementary projectors. There is only one consistency condition, which is (with $t_o = 0$)

$$\mathrm{Tr}(\bar{E}_o \rho E_o U^{-1}(t) E' U(t)) = 0, \qquad (16.6)$$

Neglecting first the correction $\delta' E$ in equation (16.5), the trace is $\mathrm{Tr}(E_o \rho \bar{E}_o)$ and it is obviously zero. The probabilities for the propositions C' and "C_o and C'" are trivially equal, from which the expected inference follows.

One must then take $\delta' E$ into account. I will consider only its effect on the consistency condition. The trace (16.6) becomes

$$\mathrm{Tr}(\rho U^{-1}(t)\,\delta' E\, U(t)\bar{E}_o). \tag{16.7}$$

One can obtain an estimate of this quantity by using a standard inequality, which is the key for all the estimates in quantum logic, namely:

$$|\mathrm{Tr}(AB)| \leq \mathrm{Tr}|A|.\|B\|. \tag{16.8}$$

The right-hand side of the inequality is the product of a trace norm and a Hilbert norm. For the Hilbert norm, one uses the part in the product (16.7) involving the projectors, namely $\|U^{-1}(t)\,\delta' E\, U(t)\bar{E}_o)\| \leq \|\delta' E\|.\|\bar{E}_o\| = \|\delta' E\|$. For the trace norm, one has simply $\mathrm{Tr}|\rho| = \mathrm{Tr}\rho = 1$ because of the positivity of ρ. The error in the consistency condition is therefore smaller than $\|\delta' E\|$, which we have seen to be exponentially small in most cases.

One can also consider determinism in the opposite time direction. It is then convenient to fix the cell C (representing, for instance, a recorded datum) and consider a cell C_o' including C_o, standing for the past to be reconstructed from the record. The reversed inference $C \Rightarrow C_o'$ can then be proved as before. (*Note*: One should not forget, however, that dissipation has not yet been taken into account and that the notion of record will be more precise when this is done.)

THE STATUS OF DETERMINISM

137. The main difference between a purely classical and a quantum conception of determinism lies in the existence of very small probabilities of error. A conditional probability underlying a deterministic logical inference is not strictly equal to 1 but only very close to 1. In the previous example, the inference $C_o \Rightarrow C'$ is controlled by the inequality

$$p((C', t)|(C_o, t_o)) \geq 1 - \|\delta' E\|. \tag{16.9}$$

This display of an error in determinism cannot be very surprising. One knows that quantum fluctuations can never be ignored, however small they may be. George Gamow illustrated them by an example in which a car crosses the walls of a garage by means of a tunnel effect. He could also have taken as an example the case of a voltmeter indicating 10 volts in place of an actual value of 110 volts because of a giant quantum fluctuation in the instrument. This sort of game is amusing but of little importance as long as the probability of the aberrations remains extremely small.

I said that the conditions for the validity of determinism are those of the Egorov theorem. There are good reasons to believe that the theorem itself is valid under wider conditions than the ones under which it has been explicitly proved. There are cases, however, when it is plainly invalid, and we saw in Chapter 11 some of the most obvious ones for which one must expect a violation of classical determinism.

An interesting case occurs when there is a potential barrier that is narrow enough to allow a tunnel effect at a macroscopic scale. For a long time, this was believed to be impossible until systems of that kind were produced in the eighties with SQUIDs (superconducting quantum interference devices). They are macroscopic superconducting rings containing a Josephson junction, and the tunnel effect is manifested by a sudden random rise in the voltage across two contacts. This is, of course, a violation of determinism that can be attributed to the extreme ingenuity and skill of the people who conceived and built it.

Another more frequent violation of Egorov's theorem occurs for a chaotic system when chaos reaches the scale of Planck's constant. An experimental test of determinism cannot, however, extend to such extreme conditions where only statistical evaluations can be used, either in the classical or in the quantum case. There are strong indications that the two statistical predictions must be very similar, in which case the difference would be only of academic interest.

THE LOGIC OF COMMON SENSE

138. When applied to physics, the logic of common sense is essentially concerned with classical properties of macroscopic objects. It permeates the human mind as a result of personal experience, of our community of culture, and through the structure of language itself. I will say nothing of other magical or mythical visions of the world because we are interested only in the ordinary version of common sense believing in causality. This means that it applies to physical situations in which determinism is valid.

The logic of common sense agrees with the quantum laws. It can be cast in the framework of the language of interpretation by reducing, at least in principle, every statement and every description pertaining to common sense to a sequence of relevant classical properties. As usual in the quantum approach, this is a history, which belongs to a formal family that one can obtain by introducing the negation of every property. Commonsense reasoning, which is in agreement with classical physics, is valid from a quantum standpoint if the family of histories is consistent and if one can prove the necessary inferences.

The essential steps of the proof have already been indicated in Section 136 and they need not be repeated. One may add that classical logic can be extended to a continuous time, in agreement with the conceptions of classical dynamics. This is because the Egorov theorem implies a correspondence between unitary quantum transformations and classical canonical transformations and because classical motion is such a group of canonical transformations. It remains to include friction effects and the vision of the world in three-dimensional space before recovering everything of common sense, but they will emerge later from the theory of decoherence.

One can therefore understand the excellent validity of commonsense logic and its extremely wide domain. These are nevertheless corrections indicating that common sense involves some errors. In ordinary circumstances, the errors are so small that they make no sense in practice except if said to be zero. When one considers, however, smaller and smaller physical systems, the probability of an error in common sense becomes higher and higher, until it is on the verge of 1, and common sense turns into nonsense. This limit occurs, of course, at an atomic level.

The resulting limitation of common sense implies drastic consequences for the philosophy of knowledge, which were stressed by Bohr, Heisenberg, and Pauli and already indicated in Chapter 5, but this topic will not be discussed now. One may, however, return to the notion of reality. One may notice that classical logic is immune to complementarity because of its restriction to classically meaningful properties. More exactly, it can ignore complementary logics because the properties it has to deal with are always mutually compatible. Technically, this means that the corresponding quantum projectors commute to a high degree of approximation. Conceptually, it means that everything we are considering at a classical level is accessible to knowledge. Every aspect we find in it can be conjugated with another. But these features are the main characteristics of our conception of ordinary reality (i.e., reality as a category): it is unique (no complementarity) and complete (everything appears as knowledgeable).

To forego these features of reality in the atomic world was certainly the most difficult *psychological* obstacle for understanding quantum mechanics. Bohr explained it again and again, and he was not always heard; sometimes he was misunderstood. To take an example: An atom is real; it exists and it is not an imagination of my mind. I cannot, however, know here and now everything I can conceive about it. This means I cannot think legitimately of anything I may imagine, nor of everything together, concerning the atom. This obstruction to my presumed intellectual universality, coming from the tiniest thing in the world, is offending, unbearable. Or is it? The pioneers of quantum mechanics, who knew the long path that had led to it, were wiser. They were aware of the huge gain for which this act of intellectual modesty is the price, a gain consisting of an immensely wider knowledge: knowing the laws, and perhaps the ultimate laws, of nature.

PHILOSOPHY OF KNOWLEDGE AND INTERPRETATION*

Because this book intends to understand quantum mechanics and not merely to justify its practical use, we may indulge in a short excursion among some philosophical questions that are certainly useful for a better appreciation of science.

Classical Physics as an Emerging Science

139.* The notion of emergence, as we are going to use it, deals with the relation between different sciences. The nineteenth century was often reductionist and assumed a hierarchy among the various branches of science. Some sciences would be reduced to and in the limit deduced from other more fundamental sciences. On the contrary, the notion of emergence (Anderson, 1972) stresses that each science has its own conceptual and methodological autonomy, while recognizing mutual relations among sciences and their global unity. Emergence may be defined as follows: A science B emerges from another science A when the basic notions and the laws entering in B can be considered as issuing from those of A (either explicitly or only from considerations of compatibility). Although the notions and the laws of B remain clearly different from those of A, they owe their efficiency to this difference.

Examples are plenty in physics. The standard model of elementary particles with its quarks, leptons, and mesons provides a foundation from which emerges nuclear physics. Each nucleon is made of three quarks, and the interactions between two nucleons are supposed (and partly proved) to follow from quark interactions, but this is of little practical avail. The relevant notions entering into nuclear physics take nucleons as their basic objects and their interaction as due to an exchange of mesons (pions, rho meson, omega meson), which the standard model would consider as made of quarks, whereas this compositeness is ignored by nuclear physics. It goes even further and still nearer to practice by disregarding meson exchange and adhering to phenomenological potentials. There are, therefore, at least two levels of emergence between fundamental particle physics and the practice of nuclear physics.

The steps from one level to the next one are elucidated by theorists, sometimes carefully (as when going from meson exchange to potentials) and sometimes sketchily (as when going from the standard model to nucleons and exchanged mesons). These links remain a topic of research, although a slowly progressing one. There are intrinsic reasons for this slowness: the difficulty of the problems is one. There is also a sociological reason: because everybody is convinced a priori of the results, the efforts for obtaining them are not really appreciated by the physics community.

Another well-known example in condensed matter physics has to do with superconductivity. The reference science A is in this case solid-state physics, which deals with electrons in a crystal lattice of ions and is still quite close to the foundations of quantum physics. The physics of superconductivity is science B. The most basic notions of superconductivity come directly from an investigation into solid-state physics. The charge of an electron induces slight deformations in the ion lattice and thereby has an action on other electrons. The consequence is an effective interaction between two electrons in the degenerate electronic gas at low temperatures, which results in the existence of strongly correlated, bound *Cooper pairs* of electrons. The emergence of the science of superconductivity consists in taking these pairs as genuine, basic objects and investigating their statistical properties (mainly their Bose condensation), their motion, and their interaction with a magnetic field. One may still reach another level of emergence when the quantum phenomena are left aside and one works directly with macroscopic properties, with their own notions and proper laws (e.g., London's law for the current or the Meissner effect in a magnetic field).

Here again, the transition from one level of emergence to the next is more or less carefully justified and remains a topic of research. One may notice that new phenomena, such as high-temperature superconductivity, provide much more incentive for research than the consistency problem between sciences A and B, whatever A and B are. One could still give many examples of emergence in physics, the relation between quantum physics and chemistry being among the most interesting ones. The vast problem of the relation between biology, chemistry, and physics stands, of course, above all others.

The reason why I opened this parenthesis is to avoid an excess of reductionism in the relation between classical and quantum physics. We just saw how the first emerges from the second, an outcome (which can probably be further improved) of the work of many people over a long time. Few, if any other, cases of emergence have been so carefully investigated and analyzed. This is probably because no other was so important yet so uncertain, because of the distance between the concepts in the two levels and the apparent opposition between determinism and probabilism.

The relation of classical physics with quantum mechanics is in that sense a remarkable case of emergence that might serve as a paradigm in the philosophy of science. It shows how the specific notions for a definite science B can be constructed *explicitly* from more general ones pertaining to a science A. The new notions are valid under special (macroscopic) conditions, where they are much more efficient. Although they are derived, they must always be preferred to the supposedly more general notions they come from, except when the correct language of interpretation is constructed and checked, that is, in the foundations of interpretation.

A Reversal of Assumptions

140.* My next remark has already been indicated in Chapter 7, and it bears on the nature of interpretation. A simple way of expressing it would be to ask the question: do we understand quantum mechanics, and if so, how?

The reflections on interpretation have always been split unequally between two great tendencies whose main leaders were Bohr and Einstein. The question of reality is certainly the borderline between their two attitudes. According to Bohr, the aim of science is only to discover the correct laws of nature, and later, should the question arise, philosophical reflection may arrive at a finer meaning and a better use of the word "reality." Einstein is more in the line of Descartes, for whom reason possesses enough power for getting directly at reality itself. This is particularly clear in the EPR paper, in which Eistein did not hesitate to define reality, or at least elements of it. To define something implies that one disposes of a deeper foundation on which the definition is built.

The majority of philosophers incline toward Einstein's approach, as do some physicists who are in search of a supposed missing piece in the quantum puzzle, for instance, hidden variables. But neither did the philosophers find their way through the new science nor the physicists reach another satisfactory theory, and no fault was found in the quantum principles.

The progress in interpretation itself were essentially always in the directions indicated by Bohr, Heisenberg, or Pauli. The results of the present chapter extend the correspondence principle, for instance. They are in conformity with Bohr's approach even though they disagree with some excessive pronouncements he made; at least the view of physics at which one arrives is very similar to Bohr's (with complementarity for instance). One may also remember that the effect of decoherence was suspected by Heisenberg, even if vaguely.

Perhaps if one were to draw a line between the Copenhagen interpretation and the present one, one could find it in a more direct reliance on basic laws. The founders still remembered a period of guesswork, of intuitive questioning, perpetually returning to solved problems. Newcomers have learned quantum mechanics at school and they can begin with it—the notion of new dwarfs standing on the shoulders of older giants remaining the traditional image of science. The principles of quantum mechanics are no longer considered to be discoveries that might still be open to improvement. They have withstood the passage of time with innumerable confirmations; therefore, the idea of relying on these principles, of searching within them for the germ of their own interpretation can take precedence over other approaches.

"To understand" (Latin *comprehendere*) means "taking something together as a whole." "To interpret" (Latin *interpretare*) means "making clear what is obscure in a text." The principles of quantum mechanics have something of an obscure text in their mathematical symbolism. Quantum mechanics itself consists of disjunctive parts (determinism versus probabilism, theory versus experiments, complementarity versus common sense) which one must comprehend as a unit. Understanding and interpreting are inseparable in that case.

When the idea of relying uniquely on the principles for interpreting and understanding was turned into a program, it was unexpectedly easy to do so. The most difficult psychological obstacle was to renounce common sense or rather, as the philosopher Edmund Husserl would have said, to put it momentarily between parentheses. We just saw in this chapter how it is recovered at the end as a gift from the theory. When trying to appreciate how this means "understanding," the whole structure of interpretation becomes clear.

Common sense has always been the basis of all attempts at understanding anything, even when distilled to philosophical principles, but these principles became an impediment when one tried to understand the minute constituents of the universe. Causality (every effect has an immediate cause) failed. Locality (every object has a definite place) had to be forsaken. There was also separability, which usually asserts that one can know all the properties of an isolated object while ignoring the rest of the universe. We saw with the EPR problem that nonseparability is the impossibility, in some circumstances, to ignore how the object was prepared and left correlated with something else now far away. Experiments confirm noncausality (in quantum jumps) as well as nonseparability; they will be considered in Chapter 22. Nonlocality is confirmed directly by interference experiments. The conclusion is obvious: the old, veneered principles of philosophy are certainly not universal.

Other, more reliable principles must be adopted. We have seen what they are: the new principles that science itself has arrived at as basic laws. The exchange between what kind of principle is taken to be first or second, premise or conclusion, is the keystone of a modern interpretation. One can comprehend as a unified whole the quantum principles, their consequences, and also common sense, if the laws of science are considered as primary. The limitations of the older principles are then found and their domain of validity is assigned. After a reversal of premise and conclusion, one understands. If deprived of this design, one still wanders in a labyrinth of misconception.

SUMMARY

The language of interpretation agrees with classical physics when dealing with the vast majority of macroscopic systems. Because it involves only classical properties, its logic is immune to the ambiguities of complementarity, and it coincides with commonsense logic, including its category of reality.

The category of causality is also valid at this level. Determinism, which is a logical equivalence between two classical properties occurring at different times, is valid. This reconciliation with quantum probabilism hinges on the idea that there is always a probability of error when determinism is involved, although this probability is negligible in most cases.

17

The Decoherence Effect

141. Two important effects are still missing from our analysis of macroscopic systems, and our purpose is now to investigate them. The first one is *dissipation*, which also goes under the names of *friction* or *damping*. The other effect is more subtle. It consists in the disappearance of quantum interferences at a macroscopic level. The foundations of physics rest on a theory in which superposition is one of its most basic features, whereas every experiment, every fact we can see at our own scale, shows no hint of this linearity. Every fact? Not even so, because of an exception in which interferences with light make the problem more mysterious: what is the peculiarity of light that explains this difference?

These problems are difficult, and a long time elapsed between the perception of their existence and the inception of their solution, whose keyword is: "irreversibility." By definition macroscopic systems have a very large number of degrees of freedom, each one of them describing a particle. We know that the particles have an erratic thermal motion and that they can draw energy from classically ordered motion at a larger scale, this effect being known as *dissipation.* I indicated in Section 61 why the environmental wave function—which is extremely sensitive to the rhythm and details of the global classical motion—loses the memory of phases. Macroscopic interferences disappear, and the phenomenon is called *decoherence*.

I gave earlier an intuitive mathematical explanation of the effect, involving directly the local phases of the total wave function. Another more precise approach will be given in the next chapter, where the essential character of decoherence appears to be irreversibility. Between these two approaches there are various models that have been essential in obtaining and understanding some important results. I will not survey these theoretical works in the present chapter, leaving them for the next one, and for simplicity will adopt a phenomenological approach: the basic results of theory will be assumed, with some essential equations they involve, and we will concentrate on their consequences in the practice of physics.

ASSUMPTIONS

142. We will use again the framework introduced in Chapter 7 (Sections 58–60). One is interested in a macroscopic system for which some collective observables have been selected. Many other degrees of freedom describe its internal environment (i.e., the inside of its constituting matter) and external environment (i.e., the atmosphere and radiation around it). We aim only to describe what can be actually observed, and this can only be the subsystem that is parametrized by collective coordinates. The word "relevant" would be perhaps better after all than "collective," although I will keep to the latter. In the case of a quantum measurement, some relevant observables are genuinely collective, and they describe the position and velocities of various parts of the apparatus and macroscopic fields, as explained in Chapters 10 and 11; but some observables pertaining to the measured microscopic system (atom, particle or else) are also relevant, if only because they are measured.

Let me define the notation. The Hilbert space in which collective observables are acting is denoted by $\mathcal{H}_c$ and the environment Hilbert space by $\mathcal{H}_e$. The full Hilbert space in which the Schrödinger equation is written is the tensor product $\mathcal{H} = \mathcal{H}_c \otimes \mathcal{H}_e$ (the notion of a tensor product meaning simply that the full wave function depends on both collective and environment coordinates). It may be convenient to think of the physical system as made of two interacting subsystems, the collective one and the environment. The total Hamiltonian will be written as

$$H = H_c + H_e + H_1, \tag{17.1}$$

where H_c is a collective operator, H_e the environment Hamiltonian (whose average value is the internal energy), and H_1 a coupling of the two subsystems. It is responsible for any energy exchange between them and therefore for dissipation.

One may notice that equation (17.1) assumes implicitly that the collective observables have been chosen once and for all. This is not always possible. In a measurement using, for instance, a Wilson chamber, a bubble chamber, or a wire chamber, new droplets, new bubbles, and new sparks are produced as time goes on. They are described by new collective observables that had no existence nor meaning beforehand. The framework of histories seems to be necessary, or at least convenient, for taking this into account, but I will abstain from entering into these intricacies.

The state of the system is described by a state operator ρ, but it brings a superfluity of good things. It contains every conceivable bit of information on the system, whether it be accessible or not, actual or a hidden mark of the past, in the minute phase variations of wave functions—too many intermingled details to be of practical use. Since one has access only to the collective observables, every available information is contained in the "collective" or "reduced" state operator

$$\rho_c = \mathrm{Tr}_e\, \rho, \tag{17.2}$$

where Tr_e denotes a partial trace on the environment.

According to quantum dynamics, the full state operator evolution is governed by the Schrödinger–von Neumann equation

$$d\rho/dt = (1/i\hbar)[H, \rho], \tag{17.3}$$

and the essential problem is to derive from this the evolution of the reduced state operator.

Note. The introduction of a time-dependent state operator differs from the point of view I used when introducing the basic principles with observables varying in time and a fixed state operator. The relation between the two approaches, however, is easy and it will be considered in Section 169.

An Example

143. Examples are innumerable but it will be convenient to give a simple, although not trivial, one. The macroscopic system is a cylinder containing a gas, the gas being the environment. In the cylinder there is a piston, with mass M and position X, which is a collective observable. This is an ideal piston, a toy for theorists, with only one degree of freedom. A spring can exert a force on the piston, deriving from a potential $V(x)$, and the collective Hamiltonian is $H_c = P^2/2M + V(X)$. If the environment were a perfect monatomic gas, its Hamiltonian would be $H_e = \Sigma_k p_k^2/2m$, that is, the kinetic energy of the atoms. As for the coupling H_1, it will be described by a model. One can assume that there is a repulsive potential U between the piston and each atom, strong enough for preventing the atoms to cross the piston, so that $H_1 = \Sigma_k U(x - x_k)$.

THE MASTER EQUATION

Preliminaries

144. When one envisions the tremendous variety of dissipation phenomena, either mechanical, solid or fluid, electrical, magnetic or chemical, and one knows, furthermore, how each of them can exhibit a

surprising subtlety when analyzed, one does not expect all of them to be explained by an explicit general theory. The same must be true for decoherence, whose field is at least as wide. The two kinds of effects are both irreversible processes, and these can be described by rather general, although not universal, theories.

Various features of dissipation also occur in decoherence. Damping is often linear in velocities, and this will be the only case of coupling with the environment one will be able to describe. For a system with only one collective degree of freedom and a kinetic energy $P^2/2M$, damping appears then in classical motion through an equation

$$\frac{dP}{dt} = F - \gamma P, \tag{17.4}$$

where F is the external force and γP a "friction force" with γ the dissipation coefficient (one speaks of *friction*, *attenuation*, *damping*, or *relaxation*: a diversity of words expressing the multiple aspects of the phenomenon).

Damping can show retardation effects, which also occur in decoherence. Retardation is familiar in electricity and electronics. In the case of an alternating current, for instance, one can use as a coordinate the charge on a condenser, with the current playing the role of velocity and dP/dt in equation (17.4) being an induction electromotive force. The coefficient γ is replaced by the imaginary part of impedance and depends on frequency. When the current varies with time in an arbitrary way, the dependence on frequency yields, after a Fourier transform, a retardation effect in dissipation. Equation (17.4) is replaced by

$$\frac{dP(t)}{dt} = F(t) - \int_{-\infty}^{t} G(t-t')P(t')\,dt', \tag{17.5}$$

showing that dissipation is not instantaneous but rather extends through time.

These characters of dissipation have been analyzed, particularly in a well-known fluctuation-dissipation theorem taking into account quantum effects and temperature (the theorem is due to Callen and Welton and to Kubo; see Landau 1967). It relates thermal and quantum fluctuations with the frequency-dependent dissipation coefficients. I will not use it even though it shows a feature we shall find again in decoherence, namely, that the results simplify considerably when the temperature is high enough.

Speaking of "temperature" would seem to contradict the fact that we are discussing irreversible processes, which are outside equilibrium. We shall see in the next chapter that "temperature" is only a convenient way for speaking of the internal energy in the environment: it is *formally* the temperature that would give the same internal energy at thermal equilibrium. As a matter of fact, the results we will obtain about decoherence do not assume thermal equilibrium (through the presence of an infinite thermostat, for instance).

I must also mention a very important point. I will denote collective observables by X, the matrix elements of ρ_c in the corresponding basis being denoted by $\langle x'|\rho_c|x\rangle = \rho_c(x', x)$. It turns out that the results of the theory depend on the choice of this basis, which cannot be arbitrary. It will be shown in the next chapter that this basis corresponds to coordinates in ordinary three-dimensional space for the various parts of the system, at least for a purely mechanical system. This may be different when electromagnetic effects are considered: for a current in a loop, X is the magnetic flux. We shall also meet the overcomplete basis of coherent radiation states in Chapter 22. The existence of a "privileged basis" in which decoherence has a simple expression is certainly not universal, and it appears to be linked with invariance properties.

A Simple Master Equation

145. The theory of decoherence and various models yield a "master equation" for the time evolution of the reduced state operator. This equation is somewhat involved, but the most important effects can still be seen in a simplified version, which is valid when the following conditions hold: (1) Retardation effects are not important. (2) The effective temperature is high enough. (3) There is a privileged basis. (4) Attention is restricted to pairs of values (x', x) for collective coordinates that are close to each other on a macroscopic scale although not necessarily on a microscopic scale. (5) One considers only for simplicity a unique collective coordinate X, as in the previous example. (6) The kinetic energy in the collective Hamiltonian H_c can be written as $P^2/2M$, where P is the canonical conjugate of X and M an inertia coefficient independent of X (M is not necessarily a mass since X is not necessarily a length).

More precision on these conditions will be given in the next chapter, but it will be enough for the present orientation to write out the corresponding master equation, which is

$$\partial\rho_c(x', x, t)/\partial t = \frac{1}{i\hbar}\langle x'|[H_c', \rho_c(t)]|x\rangle - \mu(x'-x)^2\rho_c(x', x, t)$$
$$-(\gamma/2)(x'-x)(\partial/\partial x' - \partial/\partial x)\rho_c(x', x, t). \quad (17.6)$$

The Hamiltonian H_c' in the right-hand side is not in general identical with the term H_c in equation (17.1), and it involves an action of the environment on the collective subsystem. In the previous example, this action would be due to the pressure, which is exerted by the gas on the piston. One may notice that if the first term on the right-hand side were the only one on that side, equation (17.6) would be an ordinary quantum evolution equation with the Hamiltonian H_c'. It therefore represents an ordinary collective motion.

The last term on the right represents damping. One can see it as such by considering the evolution of the average momentum $\langle P \rangle = \mathrm{Tr}(\rho_c P)$. Multiplying equation (17.6) by P and taking the trace, the contribution of the first term on the right-hand side can be identified with the average force. The second term yields zero

(even when $P = (h/i)\partial/\partial x$ acts on $(x' - x)^2$, it leaves a factor $(x' - x)$ that vanishes when the trace is taken).

When computed explicitly, the last term gives $\gamma\langle P \rangle$, as will be shown more explicitly in the next chapter. One thus obtains

$$\frac{d}{dt}\langle P \rangle = \langle F \rangle - \gamma \langle P \rangle,$$

showing that the quantity γ is the damping coefficient. The theory gives an explicit expression for it, which depends on the coupling H_1 in equation (17.1), but this expression is not very useful because generally one ignores the exact expression of the coupling. On the other hand, it is easy to find γ empirically by observation of the classical motion.

Let us now turn to the most interesting term in equation (17.6), which is second on the right-hand side and responsible for the decoherence effect. Theories give an explicit definition for the *decoherence coefficient* μ in it. Under the previous assumptions, there is a remarkably simple relation between the two coefficients for decoherence and damping, which is

$$\mu = \gamma MkT/\hbar^2, \tag{17.7}$$

where k is Boltzmann's constant and T the formal temperature representing the actual value of the internal energy. It is simply the real temperature when the environment (for instance, a gas) is not far from thermal equilibrium. This relation is extremely important since it shows that the decoherence coefficient μ is very large, because of the square of Planck's constant in the denominator. One may therefore expect that the corresponding term in equation (17.6) will dominate the evolution (except, of course, in exceptional cases when the damping coefficient γ is very small. I will come back to this point later).

Notes

It may be useful to add two remarks, one of which gives a simple argument for obtaining equation (17.7) and the other, an estimate of the orders of magnitude involved.

Justifying Equation (17.7)*

I will rely on the previous example. Assuming equation (17.6) to be correct, one can apply it when the whole system is in thermal equilibrium. H_c' is then reduced to $P^2/2M$ because in equilibrium conditions, the pressure force cancels the spring force. The reduced density operator is therefore given by the Boltzmann-Gibbs formula $\rho_c \propto \exp(-P^2/2MkT)$. When this expression is introduced in equation (17.5), the left-hand side vanishes $(\partial/\partial t \to 0)$ as well as the commutator in the first term of the right-hand side. The last two terms must therefore cancel and a straightforward calculation gives equation (17.7).

Orders of Magnitude

Notice that $\mu(x'-x)^2$ is the inverse of a time, which I will denote by t_d. Consider the case of a pendulum with a mass 10 grams and whose position is observed with a precision of 1 micron ($|x-x'| = 10^{-6}$ meter). Let the damping time of the pendulum be 1 minute ($\gamma = 1$ minute^{-1}). One obtains $t_d = 1.6 \times 10^{-26}$ second, a time so short that it is of the same order of magnitude as the time interval between two successive collisions of the pendulum with air molecules (in normal conditions). Although one is then clearly at the limit of validity of equation (17.6), the importance of the decoherence term is obvious.

DECOHERENCE

146. It is not so obvious that the second term on the right-hand side in equation (17.6) has a relation with the accumulation of fluctuations in the phase of the environment state, as I argued earlier. This may be seen however by a careful analysis of the proof of this equation or similar ones, as was done for the first time by van Kampen (1954). We shall accept this interpretation in any case and see whether it implies the decoherence effect we expect, with its suppression of interferences.

Let us consider as a first orientation the case of two coordinates x and x' with a rather large difference $(x-x')$ when compared with the quantum scale (although still macroscopically small). More precisely, the quantity $-\mu(x'-x)^2\rho_c(x',x)$ in equation (17.6) is supposed significantly larger than the matrix element of $-(i/\hbar)[H_c', \rho_c]$ expressing quantum dynamics. The damping term involving γ may be considered as negligible in view of equation (17.7). The master equation becomes then very simple:

$$(d/dt)\rho_c(x',x) = -\mu(x'-x)^2 \rho_c(x',x). \tag{17.8}$$

This is a simple differential equation whose solution is

$$\rho_c(x',x;t) = \rho_c(x',x;0)\exp\left[-\mu(x'-x)^2 t\right]. \tag{17.9}$$

When t increases, the exponent increases and the exponential decreases. The density operator becomes diagonal. But equation (17.8) is essentially the same as equation (7.4), which was proposed on intuitive grounds as the most likely answer for the paradox of Schrödinger's cat. It implies a dynamical suppression of macroscopic interferences: dynamical because it is the consequence of a coupling with the environment and because it takes a finite time.

The decoherence time is, however, extremely short because of the very large value of the decoherence coefficient. As a matter of fact, a few phonons interacting together in the matter of the object or a few external molecules or photons hitting it are enough to spoil phase coherence. These times are so small that observations made with the quickest electronic devices are always obtained too late, when decoherence has already acted and interferences have disappeared. What is thus predicted by equation (17.8) is exactly what we see in reality: no macroscopic quantum interferences.

An Example

The best example is probably the oldest one. We saw in Section 61 the case of a macroscopic pointer with an initial wave function

$$\psi(x) = c_1\phi_1(x) + c_2\phi_2(x),$$

where $\phi_1(x)$ is a narrow wave function centered at a point x_1, $\phi_2(x)$ being identical except for being centered at a point x_2 macroscopically distant from x_1. We shall assume the two constitutive wave functions narrow enough for being insensitive to decoherence (the corresponding width or uncertainty Δx being such that $\mu\,\Delta x^2\,t \ll 1$ or the limitations of decoherence to be mentioned in the next section being reached). On the contrary, the distance between the two possible positions of the pointer is macroscopic, and after a short time t, the decoherence exponent $\mu(x_1 - x_2)^2 t$ is much larger than 1 and the exponential in equation (17.8) is practically zero.

The macroscopic state operator has then become

$$\rho_c = |c_1|^2 |\phi_1\rangle\langle\phi_1| + |c_2|^2 |\phi_2\rangle\langle\phi_2|,$$

representing the random occurrence of the two positions of the pointer with the probabilities $|c_1|^2$ and $|c_2|^2$ and no quantum correlation between the two kinds of events. There is no trace left of the difficulties associated with Schrödinger's quantum cat.

The Limits of Decoherence

147. Our analysis of the equation (17.8) was rather rough, and we may wonder for which values of $|x - x'|$ it is really valid. The quantity $(i/\hbar)$ $[H_c', \rho_c]$ in equation (17.6) was neglected and is homogenous to the inverse of t. In the case of a pendulum, for instance, its matrix elements are of the order of $\omega\rho_c$, where ω is the pendulum frequency. The decoherence effect dominates as long as $|x' - x|$ is large in comparison with a distance $D = \hbar/\sqrt{MkT\omega\gamma}$. These limitations occur at a microscopic distance and they therefore have no consequence on macroscopic observations.

The exponentially decreasing behavior of the nondiagonal elements in the state operator is limited, and it will never become exactly diagonal. This is to be expected because if one were to take equation (17.9) as universal, $\rho_c(x, x', t)$ would tend toward $\rho_c(x, x, 0)\delta(x - x')$, and this would be physically absurd: a simple calculation shows that the average collective kinetic energy would become infinite. On the contrary, equation (17.6) implies a damping of the collective energy, as one can observe.

It was also noticed that there is no decoherence when dissipation vanishes. This remark is in full agreement with the most accessible occasion for seeing interferences, that offered by ordinary light. Dissipation is extremely small for a macroscopic state of light in vacuum, that is, for a state containing many photons. Just as dissipation in a gas originates in the interaction between gas molecules, dissipation in light would come only from photon-photon interactions. These interactions are not strictly zero because of higher order quantum electrodynamics effects, and Figure 17.1 shows a fourth-order Feynman graph contributing to the so-called Delbrück effect, which is photon-photon scattering. The corresponding cross section is however very small because of a factor $(e^2/\hbar c)^4 = (1/137)^4$ expressing that this is a fourth-order effect

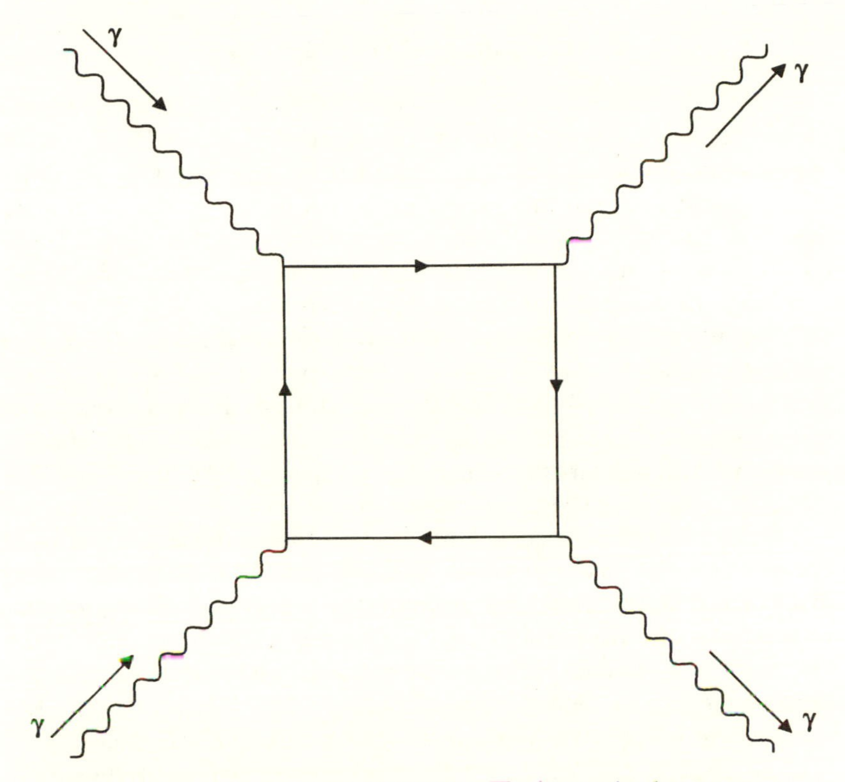

Figure 17.1. Photon-photon interaction. The interaction between two photons is governed by a Feynman graph with one electron loop. There are four electron-photon vertices, so that the photon-photon cross section is of fourth order in the fine structure constant and the interaction is therefore very small. Furthermore, kinematical factors also reduce strongly the interaction when the photon energies are much smaller than the rest energy of an electron.

(i.e., there are four electron-photon vertices in Figure 17.1). The cross section decreases rapidly when the photon energy $\hbar\omega$ is smaller than the electron rest energy mc^2, by a factor $(\hbar\omega/mc^2)^6$. In practice, the photon-photon dissipation is completely negligible, and this is why one can see optical interferences at a macroscopic level.

Other nondissipative systems can be provided by superconductors. They have been used for the construction of macroscopic systems that show practically no decoherence, behave in a quantum way, and show tunnel effects (Clarke et al., 1988).

On the Existence of Privileged Bases

148. We know that a Hermitian operator is diagonal in a unique basis of Hilbert space consisting of its eigenvectors. The tendency toward approximate diagonality of the reduced density operator in equation (17.8) cannot but raise the questions: Is this special to the coordinate basis $|x\rangle$ or not? If it is, what is so special with this basis? These questions were particularly emphasized by Wojciech Zurek (1982, 1994).

It might be recalled that among the assumptions leading to the form (17.6) of the master equation, I explicitly introduced the existence of a privileged basis. An explicit existence criterion will be given in Section 156, and it will be shown that it often follows from invariance or conservation laws. The privileged character of the collective coordinates associated with ordinary three-dimensional space will be found to follow from invariance in a change of inertial frame.

Zurek (1991) made calculations for the case I discussed briefly in Section 146. He used two Gaussian wave functions $\phi_1(x)$ and $\phi_2(x)$ macroscopically separated and found, of course, decoherence in position: namely, the diappearance of interferences when the position of the "pointer" is observed. He also noticed that when one observes the pointer momentum, interferences also tend to vanish although less rapidly than for position. These results suggest that interferences would tend to disappear between two classically observable properties, each one being represented by a rectangle of errors in phase space, as in Chapter 10. This conjecture is most probably true when the x-basis is privileged, but the question is still unsolved as of today in the general case.

SUMMARY

The decoherence effect is a loss of phase coherence in the wave functions of the environment under different macroscopic conditions. It is best expressed by a reduced density operator showing all the relevant properties of the system (involving macroscopic observables and measured microscopic observables), which is obtained by a partial trace of the full state operator on the environment.

The decoherence effect is an irreversible dynamical process. It is extremely efficient and its rate can often be simply related with a dissipation coefficient. Mathematically, at least in the main cases, it appears as an almost complete diagonalization of the reduced density operator in a specific basis. Physically, it entails the disappearance of most macroscopic interferences, thus providing an answer for the Schrödinger's cat problem.

The close relation between dissipation and decoherence implies that the only possibility of avoiding decoherence is to work with a nondissipative system, such as ordinary light or superconducting materials.

18

Theory of Decoherence

149. The aim of the present chapter is to derive from first principles the master equation (17.6) that was used in the previous chapter to show decoherence. This task cannot be done with elementary calculations, and the theoretical level of this chapter is accordingly higher than most of the rest of the book. I tried to avoid this difficulty by presenting the main results and showing their implications in the previous chapter, where the necessary physics is contained so that the present chapter can be skipped if the reader wishes to avoid too much theory.

A significant result of these investigations has already been mentioned. It consists in the existence of a privileged basis of space coordinates, at least when one can make use of a nonrelativistic change of reference frame. This is the case for a purely mechanical system and, the result is not yet fully understood. Decoherence is still a domain in which many questions remain, and many more results can be expected in the next few years.

After a long period during which attempts were made to understand a presumed universality of decoherence, attention is now switching to the artificial systems in which it takes a longer time. Such systems made the experimental observation of decoherence possible (Brune et al., 1996), as will be seen in Chapter 22. Recent interest in quantum computers could turn decoherence into a central problem in technology if the project is practicable.

In any case, I will not try here to develop the theory of decoherence in full detail. It would be too lengthy, so I will only outline the main ideas. The theoretical framework has been defined in Sections 60–62, where we saw that decoherence is basically an irreversible process. A standard method exists for approaching this kind of problem in a quantum, as well as in a classical, framework: it is known as the projection method. It is remarkably powerful, but there is an immense variety of irreversible processes, and no method is powerful enough to cover all of them. One may nevertheless expect decoherence to be both universal and extremely diversified in its detailed mechanisms, just like entropy creation is. In any case, what is presently known is far from negligible, and it is probably representative of much more general, if not universal, results.

I will conclude the chapter with some comments on the correspondence between classical and quantum physics when decoherence and friction are taken into account.

MODELS

150. The first quantitative results on decoherence were obtained with the help of models. A simple and fruitful idea was the use of the powerful theoretical tool of harmonic oscillators. It can be sometimes in agreement with the underlying physics, for instance, when phonons (elastic waves) are responsible for the interaction between the collective motion and the internal environment, as in the deformation of a wire sustaining a moving pendulum.

This model, however, is most often artificial. In the case of a piston interacting with a gas, the model would consist in replacing the gas by a collection of little springs with one end fastened on the piston and the other on the bottom of the cylinder. One may thus legitimately wonder about the generality of the results to be obtained.

The mathematical formulation of the model consists in replacing the environment Hamiltonian by a large collection of oscillators, namely

$$H_e = \sum_k (p_k^2/2m_k + m_k \omega_k^2 x_k^2).$$

with the various frequencies ω_k being different. The theory is much easier if the collection is supposed to be infinite with a continuous spectrum of frequencies. Another condition must be introduced on the coupling Hamiltonian in order not to spoil the harmonic character of the environment oscillators: the coupling H_1 must be supposed linear in the oscillators observables, that is

$$H_1 = X\left(\sum_k (\lambda_k a_k + \lambda_k^* a_k^\dagger) \right),$$

where a_k is the usual annihilation operator for the k-th oscillator and λ_k a small coupling constant.

The simplicity of oscillator dynamics is such that the calculations can be explicitly performed and one can obtain a master equation for ρ_c. The most efficient method consists in writing out the overall dynamics of the system (collective + environment) in terms of Feynman path integrals. A well-known virtue of harmonic Hamiltonians is that one can explicitly perform Feynman integrations on them through a generalization of Gaussian integrals. Here again a further condition must be added: the initial reduced density operator must be an exponential with a quadratic exponent in the oscillator variables for explicit calculations to be possible. This is realized when the initial state is in thermal equilibrium. I will not give these calculations, which are far from trivial, and only mention that they lead to a master equation quite similar to equation (17.6).

Another important model deals with an external environment. It consists in considering decoherence as an accumulation of phase changes in the wave functions of a collection of molecules or photons, a phase shift being produced when a particle scatters on a macroscopic collective object.

In spite of the difference between the two models, the master equations they yield are very similar, thus enforcing the expectation that there is universality in the decoherence effect. To be sure of this conjecture, a more general theory covering a much wider and realistic range of physical conditions is needed. The one I am going to describe is the most general one we have presently at our disposal. It contains as special cases the models that were used previously and therefore explains their agreement.

THE QUANTUM THEORY OF IRREVERSIBLE PROCESSES

151. The theory of irreversible processes originated near the end of the nineteenth century with the Boltzmann equation. Other methods in classical physics were developed for computing various kinetic coefficients for viscosity, diffusion, and thermal conductivity in a gas, for obtaining the equations of fluid dynamics on the basis of molecular dynamics, and for proceeding directly from the classical dynamics of N molecules to the Boltzmann equation. Condensed matter physics, where similar problems occur in a quantum framework, revealed a common structure in most classical and quantum calculations that had shown themselves efficient. This general formulation was called the *projection method* (Nakajima, 1958; Zwanzig, 1960, 1964; Haake, 1973).

Let us consider an isolated macroscopic system with a Hamiltonian H. Its complete state operator ρ evolves according to the Schrödinger–von Neumann equation (17.3). I will denote by A^i the "relevant" observables for which we want to know the time evolution. I already mentioned the problems one may encounter when choosing these observables, and I will only mention them to recall how much expertise helps and, more positively, that the projection method may provide a check that the choice is good, relevant, when the results are sensible and experimentally correct.

The average values of the relevant observables are denoted by $a^i(t) = \langle A^i(t) \rangle = \mathrm{Tr}(A^i(t)\rho)$. The problem is to find their time evolution. The method begins with the introduction of a "relevant" state operator depending only on the relevant observables through the expression

$$\rho_o = exp\left(-\sum_i \lambda_i A^i\right). \tag{18.1}$$

The time-dependent parameters λ_i are chosen so that the relevant state gives the same average values as the true state, that is,

$$a^i(t) = \mathrm{Tr}(A^i\rho_o(t)). \tag{18.2}$$

This starting point calls for a few comments: (1) One includes usually the Hamiltonian H as well as every conserved quantity (e.g., charge) among the relevant observables. One also includes the identity operator I so that the normalization condition $\langle I \rangle = \mathrm{Tr}\,\rho_o = 1$ belongs to the conditions (18.2). (2) The expression (18.1) of the relevant state is usually justified through considerations of information theory. The state operator minimizes the quantity of information $\mathrm{Tr}(\rho_o \mathrm{Log}\,\rho_o)$ when one can only know or accede to the relevant average values. (3) Finally, one may notice a special case when the only two relevant observables are H and I. Equation (18.1) becomes $\rho_o = \exp(-\alpha - \beta H)$, which is a state of thermal equilibrium.

152.* Before describing the method, one needs two mathematical notions:

An operator s will be called a *density* when it is of trace class (Reed and Simon, 1972, vol. 1). This means that the absolute value of s (i.e., the positive self-adjoint operator $|s|$ such that $|s|^2 = ss^\dagger$) has a finite trace. The complete and the relevant state operators are densities, but I will use densities that are not state operators (they may be non-Hermitian or nonnormalized).

One must also use the notion of *superoperators*. A convenient example is provided by the Schrödinger–von Neumann equation (17.3), which states that the matrix elements of $d\rho/dt$ depend linearly on the matrix elements of ρ, the coefficients of this relation involving the matrix elements of H. This can be interpreted by saying that a linear operator in the complex linear space of densities acts on ρ to give $d\rho/dt$. This is the definition of a superoperator. One can then write down the Schrödinger–von Neumann equation with the notation

$$d\rho/dt = \mathcal{L}\rho, \tag{18.3}$$

where the superoperator $\mathcal{L}$ is called Liouvillian.

153.* One can then formulate the projection method. It begins by introducing densities s_i that are defined by $s_i = \partial\rho_o/\partial a^i$. One can show that they possess the important orthogonality property $\mathrm{Tr}(A^i s_j) = \delta_j^i$ (the word "orthogonal" refers to the idea that the linear space of operators in Hilbert space is dual to the space of densities through the trace operation). The superoperator $\mathcal{P} = \Sigma_i s_i \otimes A^i$ turns out then to be a projector, that is, $\mathcal{P}^2 = \mathcal{P}$ (therefrom the name of the method). If one denotes by I the identity superoperator, it follows that $\mathcal{Q} = I - \mathcal{P}$ is also a projector. With these conventions, one can show that $\rho_o = \mathcal{P}\rho$ and $\rho = \rho_o + \rho_1$ with $\rho_1 = \mathcal{Q}\rho$.

Applying the two superoperators $\mathcal{P}$ and $\mathcal{Q}$ on the two sides of the Schrödinger–von Neumann equation, one obtains two coupled equations for ρ_o and ρ_1:

$$d\rho_o/dt - \mathcal{P}\mathcal{L}\mathcal{P}\rho_o - d\mathcal{P}/dt\,\rho_o = \mathcal{P}\mathcal{L}\mathcal{Q}\rho_1, \tag{18.4a}$$

$$d\rho_1/dt - \mathcal{Q}\mathcal{L}\mathcal{Q}\rho_1 = -d\mathcal{P}/dt\,\rho_o + \mathcal{Q}\mathcal{L}\mathcal{P}\rho_o. \tag{18.4b}$$

The second equation is solved formally to give ρ_1 in terms of ρ_o. This is done by writing equation (18.4b) as

$$d\rho_1/dt - \mathcal{Q}\mathcal{L}\mathcal{Q}\rho_1 = \sigma,$$

with $\sigma = -d\mathcal{P}/dt\,\rho_o + \mathcal{Q}\mathcal{L}\mathcal{P}\rho_o$. Assuming $\rho_1 = 0$ at time zero, this gives

$$\rho_1(t) = \int_0^t dt'\,\mathcal{W}(t,t')\sigma(t') \tag{18.5}$$

with

$$\mathcal{W}(t,t') = \exp[\mathcal{Q}\mathcal{L}\mathcal{Q}(t-t')]. \tag{18.6}$$

One may add a few remarks. The operator $\mathcal{P}$ depends on time as well as $\mathcal{Q}$ so that the "memory kernel" $\mathcal{W}(t,t')$ should rather be written as a time-ordered exponential of the integral of $\mathcal{Q}(t'')\,\mathcal{L}\mathcal{Q}(t'')$, with t'' going from t' to t. When one assumes that $\rho_1 = 0$ at time zero, this is tantamount supposing that the complete state operator coincides with the corresponding relevant state operator at time zero. This restriction might be removed with the burden of some nonessential complications, but I will not do so here. More important is the fact that, with equation (18.5), one introduced a definite direction of time. This does not mean that the symmetry of quantum dynamics under time reversal is broken, as is well known in statistical physics: one obtains essentially the same equation if using the opposite time direction. The existence of a privileged thermodynamical direction of time can appear only at a later stage, which is a stage of interpretation!

Bringing the expression (18.5) of $\rho_1(t)$ into equation (18.4a), one obtains a master equation for the evolution of ρ_o, which is

$$d\rho_o(t)/dt = \mathcal{P}\mathcal{L}(t)\rho_o(t) + d\,\mathcal{P}(t)/dt\,\rho_o(t)$$
$$+\int_o^t dt'\ \mathcal{P}\mathcal{L}(t)\mathcal{W}(t,t')\mathcal{Q}(t')[-d\,\mathcal{P}(t')/dt' + \mathcal{L}(t')]\,\rho_o(t'). \quad (18.7)$$

It should be stressed that equation (18.7) is still purely formal. It assumes a knowledge of the memory kernel $\mathcal{W}(t,t')$, which is as difficult to obtain exactly as would be the exact solution of the complete Schrödinger–von Neumann equation. The method will become predictive only if one is able to devise suitable approximations.

DECOHERENCE AS AN IRREVERSIBLE PROCESS

Choosing Relevant Observables

154. The method will now be applied to decoherence. The choice of relevant observables is easy since every collective observable is supposed to be relevant. It is convenient to select a basis among them, that is, a set of observables from which any other one can be obtained as a linear combination. Introducing an orthonormal basis $|x\rangle$ in the collective Hilbert space $\mathcal{H}_c$, one can use the set $|x\rangle\langle x'|$ as a basis for operators. Although it is not made of self-adjoint operators, it is equivalent to the set of observables $(|x\rangle \pm |x'\rangle)(\langle x| \pm \langle x'|)$ and $(|x\rangle \pm i|x'\rangle)(\langle x| \mp i\langle x'|)$ and I will consider the operators $|x\rangle\langle x'|$ as relevant observables according to a loose definition. When they are considered as belonging to the complete Hilbert space $\mathcal{H}$, these operators should be written as a tensor product $A^{xx'} = |x\rangle\langle x'| \otimes I_e$. The identity operator $I_c \otimes I_e$ in $\mathcal{H}$ is a linear combination of them. I will complete the list of relevant observables by adding the environment Hamiltonian H_e (which yields simpler calculations than the complete Hamiltonian H).

The average values $a^i(t) = \langle A^i(t)\rangle = \mathrm{Tr}(A^i(t)\rho)$ become, with $A^i = A^{xx'}$:

$$\langle A^{xx'}\rangle = \mathrm{Tr}(A^{xx'}\rho) = \langle x'|\mathrm{Tr}_e\,\rho|x\rangle = \langle x'|\,\rho_c|x\rangle,$$

(the notation Tr means a complete trace in $\mathcal{H}$ and Tr_e a partial trace on the environment). The matrix elements of the reduced density operator, which is precisely what we are interested in for investigating decoherence, therefore turn out to be relevant averages, and the projection method should then be appropriate for their study.

In the present case, the relevant state operator (18.1) of the projection method becomes

$$\rho_o = \rho_c \otimes \rho_e, \tag{18.8}$$

where a relevant state operator for the environment occurs and is given by

$$\rho_e = \exp(-\alpha - \beta H_e). \tag{18.9}$$

One may notice that it has the same expression as a state in thermal equilibrium. The parameters α and β are determined by the conditions $\mathrm{Tr}_e\, \rho_e = 1$ and $\mathrm{Tr}_e(H_e\, \rho_e) = \mathrm{Tr}(H_e\, \rho)$ (which is internal energy). Were the environment to consist of several different parts, there would be as many temperature parameters as there are different parts, and more generally, the method is able to cover macroscopically varying effective temperatures. I call them "effective" because they only enter in the theory as a convenience. It should be stressed emphatically that one does not assume that the environment is in thermal equilibrium. This is reflected in the fact that T is usually a fluctuating quantity.

Using Perturbation Calculus

155. The general master equation (18.7) becomes tractable when perturbation methods can be used. This happens, for instance, when the collective subsystem one is interested in is a radiation containing many photons and matter is considered as the environment. The electromagnetic coupling H_1 involves the fine structure constant, which is small enough for perturbation methods to apply.

In most cases, however, the coupling H_1 with the environment is rather strong. Solid friction, for instance, is strong, sometimes comparable to adhesive forces. When I proposed a simple model for a piston, H_1 was a potential strong enough for preventing the gas molecules from crossing the piston. Pressure is a typical expression of the interaction between a macroscopic solid and a fluid environment; confinement of a fluid by solid walls is another aspect.

The range of applicability of perturbation theory can be considerably extended if the direct macroscopic effect of the coupling is extracted from H_1. The piston example shows that the piston-gas coupling has a direct macroscopic effect, which is the pressure exerted by the gas on the piston. If one were able to extract this effect from H_1 and put it where it belongs, namely in the manifest forces included in H_c, the remaining interaction would consist only of fluctuations. It would consist of the minute changes in the piston momentum resulting from its collisions with individual molecules. This random effect is weak and, more generally, the idea I use will probably extend considerably the domain of validity of perturbation theory if it is systematically applied.

This can be done as follows: One defines a Hamiltonian ΔH_c for the average collective effect of the environment by putting $\Delta H_c = \mathrm{Tr}_e(H_1 \rho_o)$. The remaining part of the coupling, $H_1' = H_1 - \Delta H_c \otimes I_e$, will be then supposed to be small. The memory kernel in the master equation (18.7) will be developed as a perturbation series in H_1', keeping only the terms of second order in H_1' or less. The explicit calculation is too complex for being described here but the result can be. It is a master equation for the observables that are specifically relevant in the decoherence effect, that is, the reduced state operator ρ_c. It is given by

$$d\rho_c/dt + (i/\hbar)[H_c + \Delta H_c, \rho_c]$$
$$= -(1/\hbar^2)\int_o^t \mathrm{Tr}_e\{[H_1'(t), U_o(t,t')[H_1'(t'), \rho_o(t')]U_o^\dagger(t,t')]\}\, dt', \tag{18.10}$$

where $U_o(t,t') = \exp[-i(H_c + \Delta H_c + H_e)(t-t')/\hbar]$. When writing the fluctuating coupling H_1, the time dependence has been made explicit. The second term shows that ΔH_c, which is a collective operator, participates effectively in collective dynamics.

PRIVILEGED BASES

156. We saw in the previous chapter that decoherence may sometimes elect a privileged basis $|x\rangle$ where the reduced state operator shows a clear tendency toward diagonalization. This special basis is of particular importance for physics, in contrast with Dirac's often repeated statement that quantum mechanics is indifferent to the choice of a reference basis. Perhaps a few comments on that point are in order.

We saw how from its origins quantum mechanics was led toward a Hamiltonian formulation, in contrast with the much more intuitive version of mechanics we owe to Newton. Dirac made this a basic rule when he stressed the invariance of quantum mechanics under a unitary transformation (a change of basis) as the quantum equivalence of the invariance of classical Hamilton dynamics under a canonical change of variables in phase space.

Dirac was right, of course, but his pronouncement is not universal: it cannot cover interpretation. There is a technical reason for this, which is that Hamilton equations cannot account for dissipation, and we know that dissipation and decoherence go together. There is also a deeper reason. We see everyday the events of macroscopic physics happening in ordinary three-dimensional space and reminding us unceasingly that there is something very special in space coordinates, something one can lose after a beautiful canonical transformation, something specific to Newton and lacking in Hamilton: a vision in space.

One may try to follow this remark in more detail by considering the collective observables that would follow from a Newtonian approach. A physical object can be divided ideally into macroscopically small parts, which contain a very large number of atoms. Each part has a mass m, a position x (given by the coordinates of its center of mass), and a momentum p. By a comparison with Hamilton's approach, one can then distinguish two kinds of mechanical laws: (1) a relation $p = m\,dx/dt$ between momentum and velocity and (2) the fundamental law of dynamics equating dp/dt to the sum of acting forces. These forces can be extremely complicated, including external forces, interactions between neighboring pieces of matter, and frictional forces. The relation $p = m\,dx/dt$ looks very simple when compared to that, but it will reveal its importance if one asks the question: does it hold also in quantum mechanics?

The answer is affirmative. The complete kinetic energy of a little piece of matter is $\Sigma_k p_k^2/2m_k$ with an index k for each particle—electron or nucleus (we took into account the nonrelativistic character of these particles). The center-of-mass position observable is given by $X = \Sigma_k m_k x_k/m$, with $m = \Sigma_k m_k$. If one writes out the Heisenberg equation $dX/dt = (-i/\hbar)[H, X]$, one sees that the only part of H which does not commute with X is the kinetic energy of the piece of matter one is considering (the kinetic energy of another piece of matter commutes with X). The Heisenberg equation is therefore simply $dX/dt = P/m$, with $P = \Sigma_k p_k$. This implies that the relation $P = m\,dX/dt$ is exact in quantum mechanics. When X belongs to a set of collective coordinates, this relation can be written as

$$dX/dt = (-i/\hbar)[H_c + \Delta H_c, X]. \qquad (18.11)$$

Algebraic and Physical Aspects

157. Equation (18.11) does not hold in the presence of magnetic fields when the piece of matter can carry a macroscopic charge Q. The relation $md\langle X\rangle/dt = \langle P - QA(X)\rangle$ is valid as an average but is not exact among operators so that the relation (18.11) does not hold.

The exact origin of relations such as equation (18.11) is not yet completely clear. In the case where X denotes a set of collective space coordinates in the manner of Newton and when there are no magnetic fields, equation (18.11) can be shown to result from the invariance of quantum dynamics under a nonrelativistic change of reference frame.

Another important case in the theory of decoherence is concerned with a current-carrying loop, where it can be shown that equation (18.11) is satisfied if one takes for X the magnetic flux Φ through the loop. The origin of equation (18.11) is then to be found in the Maxwell equation $\operatorname{rot} E + \partial B/\partial t = 0$, and the diagonalization variable is found to be Φ.

Still another case has to do with a system of harmonic oscillators where both the environment and the collective subsystem consist of harmonic oscillators, with a bilinear coupling between the collective oscillator and its environment. One also finds in that case a convenient diagonalization basis, namely the basis of coherent (Gaussian) states. This is an overcomplete basis but it works perfectly for decoherence, at least when one is only considering a superposition of a few macroscopically different coherent states.

The situation is therefore still partly unclear. I will not try to reach a questionable universality and will restrict the present analysis to the case when the commutation relation (18.11) is valid for all the collective observables X. I will say that these observables are *microstable*, meaning by this that the classical relation between momentum and velocity remains valid, hence stable, after introducing the environment and its microscopic features.

One can give a more convenient algebraic expression of microstability. In view of the quantum definition of time derivatives, one has identically $dX/dt = (-i/\hbar)[H, X]$. Equation (18.11) therefore reduces to $[H_e + H_1', X] = 0$. The collective observable X commutes with every environment observable and therefore with H_e. The microstability condition (18.11) reduces accordingly to the simple commutation relation

$$[H_1', X] = 0. \tag{18.12}$$

This result implies a partial diagonalization of the effective coupling H_1'. Introducing a basis in the environment Hilbert space $\mathcal{H}_e$ consisting of the eigenvectors $|n\rangle$ of H_e (with energies E_n as eigenvalues), the matrix elements of H_1' in the basis $|x, n\rangle = |x\rangle|n\rangle$ become

$$\langle x, n|H_1'|x', n'\rangle = \delta(x - x')V_{nn'}(x). \tag{18.13}$$

DERIVATION OF THE MASTER EQUATION (17.6)*

Retardation Effects and Finite-Temperature Effects

158.* The master equation (17.6), which was considered in the previous chapter, can be obtained by straightforward calculations.

I give here only a few of the main steps in this derivation to call attention to some aspects of the full master equation that are not shown in the simplified expression (17.6). The most important ones are the existence of retardation effects and a rather complex dependence of the results on temperature.

Assuming the existence of a microstable basis, that is, the expression (18.13) for H_1', equation (18.10) becomes

$$\dot{\rho}_c + (i/\hbar)[H_c + \Delta H_c, \rho_c] = R_d + R_f. \qquad (18.14)$$

The two terms on the right-hand side are obtained from the first two terms of a series expansion of the evolution operator $\exp(-iH_e(t-t')/\hbar)$ in powers of $(t-t')$. The leading term R_d contains the decoherence effect. Its matrix elements are given by

$$\langle x'|R_d|x\rangle = -\int_o^t K(x',x,t-t')\rho_c(x',x,t')\,dt',$$

showing retardation effects. The kernel depends on the effective temperature, and it can be expressed in terms of the matrix elements (18.13) of H_1' by

$$K(x',x,\tau) = F(x',x',\tau) - F^*(x',x,\tau) - F(x,x',\tau) + F^*(x,x,\tau),$$

where

$$F(x',x,\tau) = \sum_{nn'} V_{nn'}(x',\tau)V_{n'n}(x,0)\exp(-\alpha-\beta E_n)\exp\{-i(E_{n'}-E_n)\tau/\hbar\}.$$

An explicit dependence on temperature is contained through the parameters α and β in these formulas. It is probably of little use, except in the rare cases when one explicitly knows H_1'. There is a similar expression for the second term R_f, which represents dissipation. There are no straightforward relations between the two kernels. They both depend on H_1', of course, reminding us of the common origin of decoherence and dissipation.

The Limit Form of the Master Equation

159. As it stands, the master equation can be used in various applications. It becomes, however, much more transparent to a physical interpretation when the simplifying assumptions I mentioned in Section 145 are introduced:

1. The retardation effects become negligible if the correlations between $V_{nn'}(x',\tau)$ and $V_{n'n}(x,0)$ differ from zero only when τ is very small. One can then smooth out the master equation over short time differences so that $F(x',x,\tau)$ is assimilated to a delta function $f(x',x)\delta(\tau)$.
2. One considers an effective temperature high enough for the factor $\beta(E_{n'}-E_n)$ to be small (the analysis of this assumption require a more detailed consideration of the function $F(x',x,\tau)$, which I will not discuss.

3. One can then introduce a restriction to macroscopically close values of x and x'. A correlation coefficient such as $f(x', x)$ has in most cases the following behavior: it is a rapidly varying function of $(x-x')^2$ and a slowly varying function of $(x+x')/2$. This dependence on $(x+x')/2$ implies an analogous dependence of the decoherence and dissipation coefficient. This is an expected result because, for instance, a friction coefficient can depend on location (think of an ice skate on asphalt or on ice). When there are several coordinates, the decoherence coefficient becomes a matrix, which is given by

$$\mu_{ij} = 2g''_{ij}((x+x')/2, 0), \tag{18.15}$$

after introducing the function $g((x+x')/2, x-x') \equiv f(x, x')$. The second derivative indicated in equation (18.15) bears on the variable $(x-x')_i$ and $(x-x')_j$.

After these simplifications, one obtains at last the master equation (17.6) in its Markovian form. As for the constitutive relation (17.7) between the coefficients of decoherence and dissipation, in the case of several variables it becomes

$$\mu_{ij} = (M\gamma)_{ij} kT/\hbar^2, \tag{18.16}$$

if one assumes that the collective kinetic energy in H_c to be written as $(1/2)(M^{-1})_{ij}P_iP_j$, the various damping coefficients being assembled in a matrix γ_{ij} giving the classical equations of motion: $dP_i/dt = F_i - \Sigma_j \gamma_{ij} P_j$.

DECOHERENCE AND CLASSICAL PHYSICS: AN OUTLOOK

160. One may hold divergent opinions on the theory of decoherence and damping. On one hand, it gives a fair idea of the physical processes that are at work and a sensible estimate for the orders of magnitude. An approach I did not describe, which uses a method of coarse graining investigated by Gell-Mann and Hartle (1993), directly relates decoherence with classical physics and seems to indicate that an extension to the physics of fluids would work. Gell-Mann (1994) has also called attention to a strong kind of decoherence, in which the wave function of the universe retains within its finest phase details the memory of all the past macroscopic and mesoscopic physical events, which would be sufficient for obtaining the kind of orthogonality one expects in decoherence.

One should mention in the list of positive results an important contribution by Caldeira and Leggett (1983a). Using the oscillator model, they investigated the evolution of the Wigner function $W(x, p)$, which is the Weyl symbol of the reduced density operator (a version of equation (10.1) where A would stand for ρ_c and $a(x, p)$ for $W(x, p)$). Quantum interferences show up through the existence of oscillations—with positive and negative values—in this function. This nonclassical behavior is particularly marked when one starts at time zero from a pure collective state involving the quantum superposition of differently located positions. In this representation, the decoherence term in the master equation becomes a diffusion in momentum cancelling out the oscillations. As time increases, the Wigner function becomes positive, and it can be identified with a classical probability distribution in phase space. Caldeira and Leggett have shown that the later evolution of this distribution is described by the classical Fokker-Planck equation. One might then start from their results to derive the classical effects usually associated with classical probabilistic effects (Brownian motion, diffusion, viscosity, fluctuations and so on).

One would then conclude that the correspondence between classical and quantum physics is essentially complete from a phenomenological standpoint.

An opposing point of view, perhaps more sensitive to mathematical rigor, would insist on serious limitations in the present theory of decoherence and dissipation. Many important dissipation effects, for instance, solid friction, are not covered by existing models nor by the theory I presented in this chapter. The theory itself is still sketchy. Its exact range of applicability has not been explored. Higher-order corrections from perturbation theory have not been estimated nor their effects investigated. I again mention the problem of choosing and classifying the collective observables. To sum up, the theory of decoherence is still in an early stage, but it can already produce many valuable results. In any case, we are now very far from the tantalizing situation that was described by Schrödinger with his fateful cat.

Comparing this situation with the theorems on correspondence that were given in Chapter 11, we see that much has been gained, at least phenomenologically, with the inclusion of friction forces and stochastic classical effects. The existence of a privileged basis, in agreement with Newton's description of macroscopic objects in ordinary space, also means that we have recovered an intuitive vision of the physics we see or touch. A mathematically inclined person would perhaps contrast the modest theoretical status of these results with the high rigor of the Egorov theorem. We can only acknowledge this difference, but mathematical perfection takes time and physicists are satisfied when they understand the physical effects at work, they know the orders of magnitude, and they see no ominous difficulty lurking. This is a fair description of the present situation.

19

Decoherence and Measurements

Two major consequences of the decoherence effect are considered in this chapter: (1) why there are no quantum interferences in a real measurement, that is, how decoherence gives an answer to the problem of Schrödinger's cat; and (2) why the rule of wave function reduction is valid for two successive real measurements. By the word "real," I mean that the measurement devices are macroscopic and that they show decoherence.

These two questions lie at the center of interpretation, and I have tried therefore to make them as clear as I could. This is why the attention is restricted to very simple examples and why the comparison between ideal von Neumann measurements (from which all the problems arose) and "real" measurements is carried on thoroughly. Later, in Chapter 21, simplicity will be replaced by generality, and the complete rules of measurement theory will be derived from the present results.

A SIMPLE EXAMPLE

161. Let us begin by introducing the specific experiment I want to discuss as a typical example of the measurement problems. It consists simply in the detection of a charged particle by a Geiger counter.

More precisely, the initial state of the particle is presumed to be in a superposition suggesting interferences. It will be assumed that the particle can be emitted along two opposite directions of space 1 and 2 and that its initial state is a pure state with the wave function $(1/\sqrt{2})(|1\rangle + |2\rangle)$ (Figure 19.1). The notation is such that $|1\rangle$ denotes the state of the particle going in direction 1.

A Geiger counter is standing in this direction 1 and the measurement is supposed to occur at a time t. When the counter detects the particle, it produces a current pulse, which is registered in a microprocessor memory. I will restrict attention to the content of this memory, which is indicated as usual by an electric voltage. Either this voltage has the value V (indicating that yes, a particle has been detected) or 0 (meaning no detection). The initial voltage is zero and it remains zero as long as there is no detection. This is what happens when the particle goes in direction 2 and does not cross the counter. In view of the simple initial

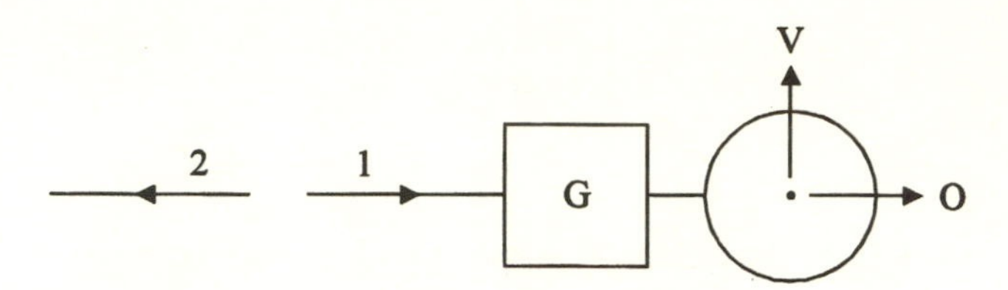

Figure 19.1. The initial state of a particle is a quantum superposition between two directions 1 and 2. A Geiger counter G (or a von Neumann ideal measuring device) is located along direction 1. It shows a datum O or V on a dial.

state, after enough time an indication 0 by the counter signals that the particle went in direction 2.

The voltage in the memory is a collective observable associated with the measuring counter, and it will be the only collective observable to be considered. My aim is to compare an ideal von Neumann measurement, from which some of the main problems of interpretation resulted, with a real measurement in which decoherence is acting. It will therefore be convenient to use similar notations for both types of measurements.

In an ideal measurement, one might use an ideal voltmeter pointer showing either the value 0 or V. These results would be associated with wave functions $|0\rangle$ and $|V\rangle$ of the device, for instance, two sharp Gaussian wave functions of a voltage variable v respectively centered at $v = 0$ and $v = V$. I will assume that the same wave functions can be obtained for the collective potential observable in the case of a real Geiger counter. This representation is, of course, rather far from the classical properties we considered in Chapter 10, but it can only favor the quantum features of measurement data, which are the origin of our difficulties and which we wish to remove.

AN IDEAL MEASUREMENT

162. The theory of an ideal measurement has been given in Section 49 and I will therefore be brief, except for again stressing its difficulties.

The reduced density operator ρ_c of the measuring device can be introduced for an ideal measurement as well as for a real one. Its use is justified in the first case by the fact that one learns of a measurement result by looking at the pointer and not at the particle. The trace defining ρ_c in terms of the full density operator ρ must then be performed on the particle degrees of freedom.

At time t, one has $\rho = |\psi_t\rangle\langle\psi_t|$, with

$$|\psi\rangle = \frac{1}{\sqrt{2}}(|1\rangle \otimes |V\rangle + |2\rangle \otimes |0\rangle).$$

Since the two particle states $|1\rangle$ and $|2\rangle$ are orthogonal, the partial trace on the particle degrees of freedom is the sum $\langle 1|\rho|1\rangle + \langle 2|\rho|2\rangle$ and the reduced state operator is

$$\rho_c = \tfrac{1}{2}(|0\rangle\langle 0| + |V\rangle\langle V|). \tag{19.1}$$

This result may be puzzling. It looks the same as the one we expect from decoherence and there is certainly a difference somewhere. The measured particle is not lost, however, in an ideal measurement. Let us therefore devise a magnetic guide as in Figure 19.2, which will bring the two particle states 1 and 2 along the same direction after time t. Whether or not the particle crossed the counter, at time t' it is in the same state $|3\rangle$, and the full wave function for the particle and the counter is at that time

$$|\psi'\rangle = \frac{1}{\sqrt{2}}(|3\rangle \otimes |V\rangle + |3\rangle \otimes |0\rangle).$$

The reduced density operator for the counter is then given by

$$\rho_c = \tfrac{1}{2}(|0\rangle + |V\rangle)(\langle 0| + \langle V|). \tag{19.2}$$

The crossed terms $|V\rangle\langle 0|$ and $|V\rangle\langle 0|$ allowing interference effects have reappeared in the state of the counter. They do not affect the probability for seeing the pointer in position 0 or V but they can be exhibited if one looks at the "switching" observable $S = |0\rangle\langle V| + |V\rangle\langle 0|$. Its average value is zero for the mixed state (19.1) and 1 for the pure state (19.2), thus providing a test for the survival of interferences.

This shows the most problematic aspect of an ideal measurement: the data it yields are not obtained once and for all. Apparently lost interferences can be regenerated later in the measuring device by an action on a distant system (the particle). There is no possibility for considering experimental facts as being firmly established. One may see the result as a particularly vicious consequence of EPR correlations or express it by saying that Schrödinger's cat cannot be dead once and for all, because evidence for his survival can always be retrieved.

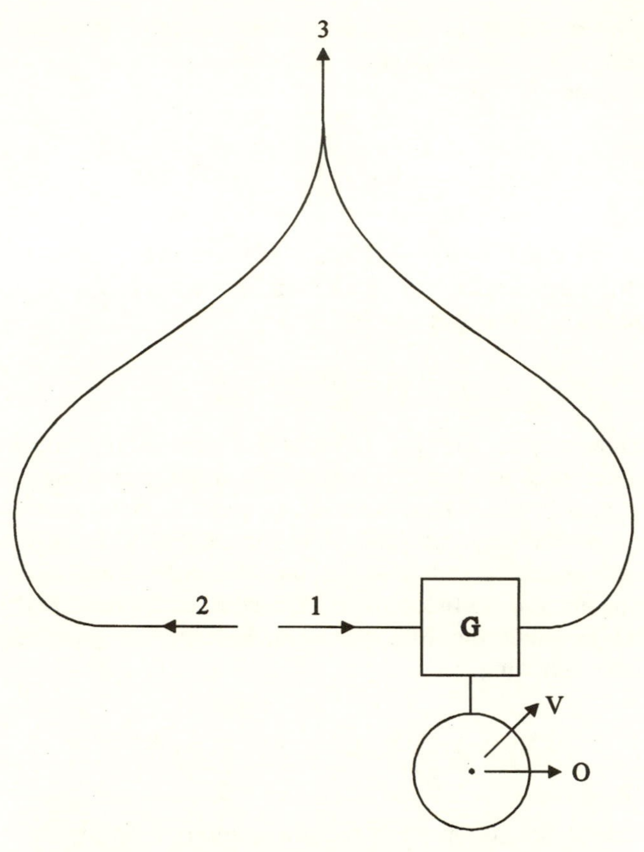

Figure 19.2. A von Neumann ideal measurement does not produce a factual datum. The particle shown in Figure 19.1 is brought along direction 3, whatever its initial direction was. When the two paths join together, the state of the measuring device becomes a pure state, with no memory of a previous mixed state where the result of a measurement could be read.

A MEASUREMENT WITH DECOHERENCE

163. In keeping with my intention of maintaining a close relation between ideal and real measurements, I still consider the same two states $|0\rangle$ and $|V\rangle$ for the memory voltage, except that this is now a *collective* observable and there is an environment in which decoherence occurs. The state $|V\rangle$, for instance, is represented by a wave function of the voltage variable v centered at $v = V$ and we assume it to be sharp

enough so that it is not affected by decoherence during the lapse of time I consider. The difference V between the two voltage data is, on the contrary, macroscopic.

Let t again be the time at which the particle crosses the counter, and for simplicity, assume it to be sharply defined. Decoherence, although very rapid, begins immediately after. Applying the results in Section 147, the reduced state operator for the memory is given after a time lapse Δt following t by equation (17.8), that is,

$$\rho_c(t+\Delta t) = \tfrac{1}{2}(|0\rangle\langle 0| + |V\rangle\langle V|) + \tfrac{1}{2}(|0\rangle\langle V| + |V\rangle\langle 0|)\exp(-\mu V^2 \Delta t), \qquad (19.3)$$

where μ is the decoherence coefficient.

*An Estimate of the Decoherence Coefficient**

One can take the evaluation of this coefficient as an occasion for an exercise in decoherence theory. Let one consider the memory as an (L, R, C) electrical circuit where L is a self-inductance, R a resistance, and C a capacity. Let the charge on the capacity be considered as a variable (replacing a position coordinate) and the current as the corresponding velocity. The role of the inertial mass M is taken by L and the damping coefficient γ is equal to R/L (R is more precisely the imaginary part of impedance). The usual decoherence exponent $MkT(x-x')^2\gamma\,\Delta t/\hbar^2$ is replaced by $LkT(q-q')^2(R/L)\,\Delta t/\hbar^2$, which is equal to $RkTC^2(v-v')^2\,\Delta t/\hbar^2$ after taking account of $q = Cv$ under normal circumstances, where the capacity dominates in a memory. One can replace $v - v'$ by V when considering the difference between the two states, and equation (18.3) then becomes

$$\rho_c(t+\Delta t) = \tfrac{1}{2}(|0\rangle\langle 0| + |V\rangle\langle V|) + \tfrac{1}{2}(|0\rangle\langle V| + |V\rangle\langle 0|)\exp(-RkTC^2V^2\Delta t/\hbar^2). \quad (19.4)$$

After a short time, the exponential is negligible and the state operator is given by equation (19.1).

One may then consider what happens if the particle survives the measurement and is brought again to a unique state $|3\rangle$ at a time t' after the measurement, whatever the result may be. This is of no consequence. The partial trace yielding $\rho_c(t')$ must be performed on the environment and the particle together. Intuitively, the orthogonality of the environment wave functions associated with the results 0 and V is enough for giving to $\rho_c(t')$ the diagonal form (19.1). A more elaborate analysis using the theory given in the previous chapter yields the same result (the particle is then simply considered as belonging to the environment).

The main difficulty of quantum mechanics therefore seems to have disappeared: not only does one eliminate macroscopic quantum interferences but also, as a consequence, the measurement data are cleanly separated once and for all.

Decoherence versus Pure States

164. I would like to add a few comments on the significance of decoherence in measurement theory. The following criticism is often made of the meaning of decoherence: If a system is globally in a pure initial state, its evolution according to the Schrödinger equation will maintain it in a pure state. Decoherence is therefore not a fundamental effect but only an appearance in macroscopic objects due to a willful ignorance of anything but collective observables.

The usual answer is that one can only acknowledge phenomena, which are described by these collective observables. There is no meaning in discussing as belonging to physics a fine property of a global state that will never be reached experimentally. One may notice an interesting divergence between the two approaches. The first one carries a memory of Einstein's touch, with some realism (the state exists and one can therefore speak of it without knowing it) and some pragmatism (questions that cannot be submitted to any experiment do not make sense). The second approach is more in the spirit of Bohr with his insistence on truth, which can be only settled by observed classical phenomena. John Bell was a proponent of the first viewpoint, although he also proved himself a very pragmatic physicist by granting that decoherence can solve the main problems of measurement theory "for all practical purposes." The question is then to decide whether physics may have or should have purposes other than practical ones.

Bell considered also the same question from a more experimental standpoint (Bell, 1975; see also d'Espagnat, 1994). We have seen that an action on the measured particle cannot affect the state of the apparatus; however, one could also contemplate an action on the apparatus itself. Could one perform a measurement on the environment and reveal the persistence of quantum interferences that were believed to be destroyed? I will return to this important question in the next chapter.

On the Continuous Action of Decoherence

165. One also may add a practical comment on the action of decoherence. We assumed that one can start at a sharp time t with a

superposed state of the particle and the counter, decoherence acting afterward. This is convenient for applying simply the theory of decoherence as we know it, but one may wonder whether this clear-cut superposition is ever produced. If not, how would this affect our conclusions?

One can look more closely at a real Geiger counter. Its active part consists in a dielectric medium that is maintained on the verge of electric breakdown by a condenser. When a charged particle enters the medium, it ionizes a few atoms on its way and produces some free electrons. Those are accelerated by the electric field and they gain enough energy for ionizing other atoms. New electrons ionize more atoms, and there is soon a large enough number of free electrons to carry a current between the condenser plates, resulting in a spark. The pulse of current can be registered by an electronic device.

Leaving the electronic device aside, one sees that the main effects occur inside the dielectric medium, that is, in the environment. As soon as one free electron and one ion are produced, they interact with their neighbors and a decoherence effect begins, weak at first but then rapidly significant and further reinforced by other ionizations. As soon as a few electrons have been liberated, one can consider as a matter of fact that decoherence is already there. Were one to discharge the condenser at this early stage, that decoherence would be nevertheless fully effective although no macroscopic phenomenon would signal it (this is the microscopic "strong" decoherence mentioned by Gell-Mann [1994]).

One can therefore assert that a quantum superposition of macroscopic states is never produced in reality. Decoherence is waiting to destroy them before they can occur. The same is true for Schrödinger's cat: the stakes are put down as soon as a decay has been detected by the devilish device, before the poison phial is broken. The cat is only a wretched spectator.

SUCCESSIVE MEASUREMENTS

An Experiment

166. Turning now to the problem of wave function reduction, I will consider two successive spin measurements of a spin-1/2 atom. It is initially prepared in a pure state $S_x = 1/2$ and first enters a Stern-Gerlach device oriented along the z-direction (Figure 19.3). The magnetic field and the magnet producing it are considered as part of the conditioning apparatus, and we saw earlier how these can be replaced by a classical field producing different atom trajectories according to the value of S_z. One needs also a measuring device, and in the spirit of

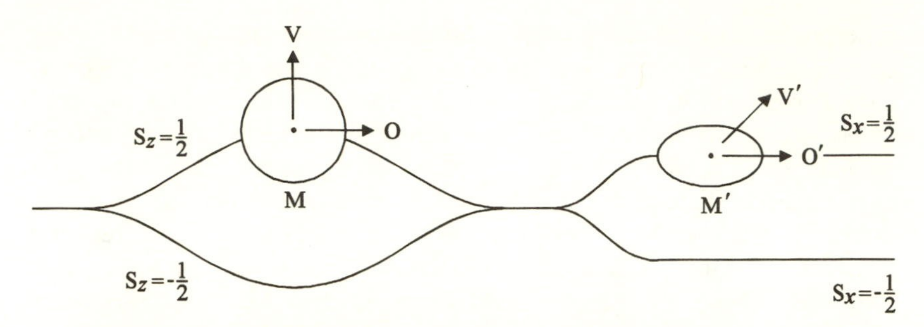

Figure 19.3. Two successive measurements. A first measuring device M measures the z-component of a particle spin. The corresponding datum is either O or V. A second device M' measures the x-component of spin afterward and the datum is either O' or V'.

our previous discussion, I will assume that a detector M stands along the $S_z = +1/2$ trajectory. It shows a voltage V in its memory when it detects and a voltage 0 when it does not (in which case the atom has gone through the $S_z = -1/2$ trajectory).

I assume with Feynman (Feynman, Leighton, and Sands, 1965) that whatever its initial trajectory, the atom enters a second Stern-Gerlach field oriented along the x-direction. There is a detector M' standing on the $S_x = +1/2$ outgoing trajectory, and after measurement its memory will show either a voltage V' or $0'$ (where the prime is intended to avoid confusion with the reading of the first device M).

I will again proceed by a comparison between the cases when the detectors are ideal and when they are real. In the latter case, one must take into account the existence of an environment e for the detector M and another e' for M'. It will be convenient to consider that the second detector M' is always real although the fixation of its data by decoherence and its environment e' will play a minor role in the discussion.

It will be convenient to use a notation indicating both the state of the atom and of the detectors; for instance, I denote by $|x+\rangle$ and $|x-\rangle$ the eigenstates of S_x with eigenvalues $+1/2$ and $-1/2$. The two states of M (either ideal or collective) are denoted as before by $|0\rangle$ and $|V\rangle$, those of M' being $|0'\rangle$ and $|V'\rangle$. The notation will be simplified further, for example, by substituting $|x+, V, 0'\rangle$ for $|x+\rangle \otimes |V\rangle \otimes |0'\rangle$. The notation for projectors will also be condensed: The property of the memory of M showing a voltage V is denoted by the projector $|V\rangle\langle V|$ whereas its full expression is $I_Q \otimes |V\rangle\langle V| \otimes I_{M'}$ in the ideal case and $I_Q \otimes |V\rangle\langle V| \otimes I_e \otimes I_{M'}$ in the realistic one.

The unfolding of events is then as follows:

- At the initial time zero, the state of the whole system is $|x+,0,0'\rangle$ (I do not write out the state of the environment, which would need state operators).
- At time t_1, the atom interacts (or not) with the detector M, which indicates then a voltage V (for the atom state $z+$) or 0 (for $z-$).
- At a time t'_1 soon after (in the realistic case), decoherence is complete in M.
- At time t_2 (later than t'_1), the atom interacts (or not) with the detector M', which shows either a voltage V (for $x+$) or 0 (for $x-$). I will not explicitly discuss later decoherence in M'.

Ideal Measurement

167. Let us first consider the case when M is an ideal von Neumann detector. The language of histories will be useful and I will rely on it. I consider a complete family of histories, including the possible results of the measurements, with the projectors $|V\rangle\langle V|$ and $|0\rangle\langle 0|$ at time t_1, $|V'\rangle\langle V'|$ and $|0'\rangle\langle 0'|$ at time t_2.

Considering the history a, where the first and second results are V and V', the corresponding history operator showing explicitly time evolution is

$$C_a = U^\dagger(t_2,0)|V'\rangle\langle V'|U(t_2,t_1)|V\rangle\langle V|U(t_1,0). \qquad (19.5)$$

I introduced a time evolution operator $U(t,t')$ rather than $U(t-t')$ in view of the general argument given in Section 114. This is justified in the ideal case since a von Neumann measurement occurs at a definite time because of a time-dependent interaction between the atom and the detector.

The family of histories is of the type (two times, yes-no) already considered in Section 116, and there is only one consistency condition, which is

$$\mathrm{Tr}(C_a\,\rho(0)C_b^\dagger) = 0, \qquad (19.6)$$

the history b corresponding to the two successive results 0 and V'.

This condition contains most of the physics and the logic we need, and one must therefore analyze it carefully.

168. Leaving aside the environment of M', $\rho(0)$ in the ideal case can be replaced by $|x+,0,0'\rangle\langle x+,0,0'|$ and equation (19.6) becomes

$$\langle x+,0,0'|C_b^\dagger C_a|x+,0,0'\rangle = 0. \qquad (19.7)$$

Let us follow the action of the various terms in $C_a|x+,0,0'\rangle$. First of all, $U(t_1-\varepsilon,0)|x+,0,0'\rangle = |x+,0,0'\rangle$ where ε is a very short time; therefore, we are considering the state immediately before the first measurement (we do not have to take the atom space motion into account explicitly, although that would be easy, but at the price of more complex mathematical formulas).

The first detection occurs between the times $t_1-\varepsilon$ and t_1. Its effect is simple in the z-basis where one has $U(t_1,t_1-\varepsilon,)|z+,0,0'\rangle = |z+,V,0'\rangle$ and $U(t_1,t_1-\varepsilon,)|z-,0,0'\rangle = |z-,0,0'\rangle$. The action of the first two operators in C_a—see equation (19.5)—yields $|V\rangle\langle V|U(t_1,0)|x+,0,0'\rangle = 2^{-1/2}|z+,V,0'\rangle$, where one used $|x+\rangle = 2^{-1/2}(|z+\rangle + |z-\rangle)$. One may notice for later use that the same result would be obtained if in place of the projector $|V\rangle\langle V|$ representing the "datum" from the first detector, the projector $|z+\rangle\langle z+|$ representing the corresponding spin property was used. This equivalence between a datum and a microscopic property ("result") will be found to be increasingly significant as measurement theory is developed.

The next terms in C_a are obtained exactly in the same way, and after a straightforward calculation they yield

$$|V'\rangle\langle V'|U(t_2,t_1)|V\rangle\langle V|U(t_1,0)|x+,0,0'\rangle = (1/2)|x+,V,V'\rangle.$$

It can again be noticed that the "datum"projector $|V'\rangle\langle V'|$ for the second measurement could have been replaced by the "result" projector $|x+\rangle\langle x+|$. In the case of history b, one would obtain

$$|V'\rangle\langle V'|U(t_2,t_1)|0\rangle\langle 0|U(t_1,0)|x+,0,0'\rangle = (1/2)|x+,V,V'\rangle.$$

Multiplying the two factors $U^\dagger(t_2,0)$ and $U(t_2,0)$ together give the unit operator in the product $C_b^\dagger C_a$ and the trace in the consistency condition, rather than satisfying condition (19.6), is given by

$$\mathrm{Tr}(C_a\,\rho(0)C_b^\dagger) = 1/4.$$

One may notice that a direct calculation with the spin projectors without any mention of the detectors (as in Section 119) would have given the same result.

The conclusion is drastic: nothing sensible can be drawn from an examination of the two so-called measurements because there is no reliable logic for dealing with them.

REAL MEASUREMENTS

I will now compare the previous negative results with what happens in a real measurement (the notion of a "real" measurement meaning again that decoherence is taking place).

Decoherence and Histories

169. The theory of decoherence given in Chapters 17 and 18 was based on a time-varying reduced density operator, whereas histories rely on time-varying observables. This difference will give me an opportunity to show how the switch from one framework (Schrödinger's) to the other (Heisenberg's) can be made in a simple case.

The detector M is now described realistically by taking decoherence into account. We are interested in its possible data, which are fixed only at time t_1' by decoherence as observable phenomena, decoherence being at work between times t_1 and t_1'. These data are the properties I introduce in the histories. I will begin by indicating some slight changes in the previous analysis, that are rather trivial and may be omitted in a first reading.

The operator C_a becomes $U^\dagger(t_2,0)|V'\rangle\langle V'|U(t_2,t_1')|V\rangle\langle V|U(t_1',0)$. The operator $U(t_1',0)$ is split into $U(t_1',t_1)U(t_1,t_1-\varepsilon)U(t_1-\varepsilon,0)$. One has again $U(t_1-\varepsilon,0)|x+\rangle=|x+\rangle$ and $U(t_1,t_1-\varepsilon)U(t_1-\varepsilon,0)|x+,0,0'\rangle = 2^{-1/2}(|z+,V,0'\rangle+|z-,0,0'\rangle)$. One is thus left with the calculation of

$$\begin{aligned}&\tfrac{1}{2}\mathrm{Tr}\{|V'\rangle\langle V'|U(t_2,t_1')|V\rangle\langle V|U(t_1',t_1)\\&\quad\times(|z+,V,0'\rangle+|z-,0,0'\rangle)\rho_e(\langle z+,V,0'|+\langle z-,0,0'|).\\&\quad\times U^\dagger(t_1',t_1)|0\rangle\langle 0|U^\dagger(t_2,t_1')\},\end{aligned}$$

which should be zero if the consistency condition is valid. This is a bit involved but mostly easy. The action of $U(t_2,t_1')$ is evaluated as before after a short stop before the second measurement and gives $|x+,V,0'\rangle \to |x+,V,V'\rangle$ as well as $|x+,0,0'\rangle\to|x+,0,V'\rangle$, $|x-,V,0'\rangle\to |x-,V,V'\rangle$, and $|x-,0,0'\rangle\to|x-,V,V'\rangle$. The only non-trivial, though easy, problem is to write down the action of $U(t_1',t_1)$ where decoherence enters.

One can introduce the quantity

$$F(t_1')=\tfrac{1}{2}\mathrm{Tr}_e\{U(t_1',t_1)(|z+,V,0'\rangle\langle z+,0,0'|)\otimes\rho_e(t_1)U^\dagger(t_1',t_1)\},$$

which is an operator in the Hilbert space of the system *atom* $+M+M'$. The consistency condition is then

$$\mathrm{Tr}\{|V'\rangle\langle V'|U(t_2,t_1')F(t_1')U^\dagger(t_2,t_1')\}=0, \tag{19.8}$$

where the trace does not involve the environment e.

The problem hinges essentially on the calculation of $F(t_1')$. I will encourage the reader at this point by saying that everything about wave function reduction will result from this calculation once and for all.

One introduces a time-dependent operator $F(t)$, where t varies continuously after replacing t'_1 by t in equation (19.7) with $t_1 \leq t \leq t'_1$. One may then return to the proof of the master equation for decoherence in Chapter 18, but because this proof was classified as not essential to a general reading, I will mention only that nowhere in the proof did one use the fact that the reduced operator in it is a state operator (it need only be a density in the sense given in Section 152). One may consider therefore the operator

$$(|z+,V,0'\rangle\langle z+,0,0'|)\otimes\rho_e(t_1)$$

as if it were a genuine initial state operator. Decoherence states that as time goes on, $F(t)$ is a nondiagonal element in the privileged basis $|z\pm,v,v'\rangle$ and therefore must vanish very rapidly.

The main answer is now obtained: our description of the two successive measurements makes sense, and we have only to draw its consequences.

THE REDUCTION FORMULA

170. One needs now only to conclude the calculation by obtaining the probability for the various relevant histories.

Consider for instance the history a. Its probability is given by

$$p(a)=\mathrm{Tr}\{|V'\rangle\langle V'|U(t_2,t'_1)G(t'_1)U^\dagger(t_2,t'_1)\},\tag{19.9}$$

where

$$G(t'_1)=\mathrm{Tr}_e\{U(t'_1,t_1)(\tfrac{1}{2}|z+,V,0'\rangle\langle z+,V,0'|\otimes\rho_e(t_1))U^\dagger(t'_1,t_1)\}.$$

The same analysis of decoherence gives the result

$$G(t'_1)=\tfrac{1}{2}|z+,V,0'\rangle\langle z+,0,0'|.$$

The only effect of the evolution operator $U(t_2,t'_1)$ is to represent the measurement at time t_2 by

$$|V'\rangle\langle V'|U(t_2,t'_1)|z+,V,0'\rangle=\langle x+|z+\rangle|x+,V,V'\rangle,$$

and as previously done, one can replace the datum projector $|V'\rangle\langle V'|$ by the projector $|x+\rangle\langle x+|$ for the corresponding value of spin. The trace over M and M' is trivial and one obtains

$$p(a)=\tfrac{1}{2}|\langle x+|z+\rangle|^2.\tag{19.10}$$

It turns out that the well-known formula usually derived from the unnecessary hypothesis of wave function reduction is contained in the result (19.10). There is, however, an inconvenience: the simplicity of the example we considered gives results from which a general conclusion is not obvious. One should therefore extend this example to obtain a more instructive one.

Rather than a spin-1/2 system, let us consider another system with an arbitrarily large Hilbert space. The initial state of the measured system is supposed to be a pure state $|\psi\rangle$. In place of S_z, one measures an observable A. In order to avoid unnecessary complication, I will assume the eigenvalues a_n of A to be discrete and nondegenerate. The detector M indicates a datum V_n for the eigenstate $|a_n\rangle$. As before, I also assume that the measured system remains in the state $|a_n\rangle$ after the measurement if it entered in that state. These are the so-called measurements of Type I (in Pauli's sense), also called "ideal" measurements, although in a sense different from the one we used in this chapter (more general results will be given in Chapter 21).

Similarly, the second measuring device measures an observable A' with nondegenerate eigenvalues a'_m associated with data V'_m. Everything in the previous analysis remains essentially the same. There are more histories, more consistency conditions that are verified and more probabilities, but the calculation is unchanged: one needs only to insert the indices n and m here and there.

One can then obtain a more transparent expression for the probabilities when considering a history in which the two successive data are V_n and V'_m (the associated results being a_n and a'_m). Equation (19.10) is then replaced by

$$p(n,m) = |\langle a_n|\psi(t_1)\rangle|^2 . |\langle a'_m|U(t_2,t'_1)|a_n\rangle|^2 . \tag{19.11}$$

This is exactly the well-known reduction formula. The probability for obtaining the result a_n in the first measurement occurring at time t_1 is given by Born's formula $|\langle a_n|\psi(t_1)\rangle|^2$, and the conditional probability for obtaining the result a'_m hereafter from the second measurement at time t_2 is given as if the state resulting from the first measurement were $|a_n\rangle$ at time t'_1 and evolved in between according to the Schrödinger evolution operator $U(t_2,t'_1)$.

This is of course a fundamental result, which will be reconsidered, extended, and commented upon in Chapter 21.

SUMMARY

The behavior of a macroscopic measuring device in which decoherence takes place differs drastically from the behavior of a theoretical von Neumann model: a datum cannot be changed or contradicted by a later external action or evolution.

In the case of two successive measurements, one can compute the probabilities of the expected results for the second measurement when a datum has been acquired by the first one. This is given by the formula that was based traditionally on wave function reduction. There is, however, no specific reduction effect affecting the measured system, and the result is a consequence of the decoherence effect in the first measuring device.

20

Fundamental Questions

The so-called "fundamental" questions in the present chapter have mostly a flavor of epistemology. Some people (of whom I am one) believe them to be essential. Readers should be advised however that these questions have no practical (experimental) consequences, as far as one knows. So, pragmatists may skip this chapter.

171. "Our reason has this peculiar fate that...it is always troubled with questions that cannot be ignored, because they spring from the very nature of reason, and which cannot be answered, because they transcend the powers of human reason" (Kant). If not a definition, this well-known sentence provides a good description of the "fundamental" questions to be considered in this chapter, however, with a difference: it seems that in physics some questions can be answered.

Which are these questions? Some of them have already been discussed: causality, for instance. Is quantum physics a description of reality? (I have touched on that one.) Why is there a direction of time? What is a fact? Is there a reduction of the wave function? How can one understand that a unique result occurs at the end of a measurement?

There are massive books and even good books on these matters. I will not try to give final answers (it would be preposterous) but rather try to show why most of the questions admit simple answers, even if they are not universally accepted. They are pragmatic answers, expressing that no real disagreement between theory and observation can be found anywhere. Who asks for more must therefore enter into philosophy, where experience cannot decide (except, of course, if physics itself changes and goes beyond our expectations).

I will try to be brief. There are, however, two very Kantian issues that must be put forward and stand as preliminary to all others: the relation between mathematics and physical reality on one hand and the meaning of very small probabilities on the other. It seems that they cannot be resolved because they transcend the powers of human reason.

The mystery of a correspondence between a mathematical construct and the multitude of scientific facts remains as great as ever. Can one say for instance, as some people do, that there exists a monstrous wave function of the universe? But in what sense? Does it exist as an honest mathematical function we play with or is there something like it existing in another guise? The last assumption looks to me like an outrageous excess in metaphysics. To find on the other hand why physics obeys quantum laws—"Why the quantum?" as John Wheeler says—or to understand deeply the nature of mathematics goes beyond our powers.

The controversy between pragmatists and fundamentalists in the field of quantum physics often hinges on the meaning of very small probabilities. For a pragmatist, a probability of 10^{-100} is identical to another of $10^{-1,000,000}$: the probabilities are zero. For some fundamentalists, these small probabilities are worth being considered, and absolute truth means probability absolutely equal to one. Fortunately (for the pragmatist) there is a science still more fundamental than the one I discuss. Quantum physics uses probabilities, and therefore, in some sense, probability calculus is more fundamental. What do specialists say about it?

The most thorough analysis of this question, at least to my knowledge, was made by the mathematician Émile Borel, one of the founders of modern probability theory (Borel, 1937, 1941). When considering the applications of this science, he proposed a unique and sweeping axiom: one must consider that events with too small a probability never occur. The main points are that (1) such events are irreproducible, (2) they have no statistical sense, and (3) they cannot belong to an experimental science (a definition of a "too small" probably will be given later). This will be the foundation of many of my statements.

Finally, it should be mentioned that when the word "interpretation" is used in the present chapter, it refers to the one I deal with in this book.

REALISM

172. We shall proceed mostly by questions and answers.

Question. What do we know of reality from a knowledge of quantum mechanics?

Answer. There are two different assumptions in realism: (1) Something exists outside our mind. (2) It can be known. The first statement is considered as obvious by any physicist. The second one has received so many different meanings, particularly after the advent of quantum mechanics, that tomes with no clear conclusion have been written on it. I don't know what "to know" means when used with such a wide extension.

People who believe in a wave function of the universe have given a curious twist to the second assumption of realism: This wave function is "something" containing a huge amount of knowledge, a total knowledge which is, however, inaccessible to our experimental means. This looks like a most extreme form of Platonism: there would be an Idea, called *psi*, more real than the reality in our own cave. What mathematics can lead some people to dream of is strange....

Leaving this aspect aside and returning to the question itself, one may remenber that a question has a meaning only if it can be stated in a definite logical framework. The question about realism refers to conventional wisdom (common sense), and we saw that common sense can receive a different acceptation when recast in the mould of quantum principles. So, we know much about the reality of atoms and particles, but what we know is not unique, not factual. It can dissolve in complementarity although complementarity is not negation. Whether this is realism or not is a matter of definition: of language.

THE DIRECTION OF TIME

173. *Question.* There is no privileged direction of time in quantum dynamics (except for a tiny *CP* violating term). Why, then, do we see one?

Answer. We know of four privileged directions of time. There is one in cosmology: the direction in which our universe is presently expanding rather than contracting. I will leave it aside. Then there is the well-known direction in thermodynamics, to which one must add the direction of decoherence. Finally, there is a direction of time in logic.

Let me elaborate on the latter. I will insist once again on the fact that any discussion, any explanation, must rest on a language endowed with logic. The only reliable one we know of uses quantum histories. One may remember that the properties in a history were given in a definite order: the time order of their occurrence. One cannot logically combine two families of histories with opposite directions of time, so one direction must therefore be chosen once and for all. Of course, there are two possible directions with no apparent distinction except for their incompatibility.

We know that the two most important ingredients in the proof of logical consistency are classical physics and decoherence. Classical physics, as long as it does not include dissipation, maintains an equivalence between the two directions of time. But a definite one must be chosen for decoherence, the same as in thermodynamics. It turns out that decoherence implies the logical consistency of some histories, under the condition that the same direction of time is used for both logic and decoherence. Histories proceeding in the opposite time direction, like a motion picture running backward, are not consistent (at least the ones relying on decoherence are not, and we saw that all of them must have the same direction).

Then comes the relation between a logico-mathematical theory and physical reality: one chooses, of course, the direction in which the mathematical parameter t is running as representing the one we see in nature.

TIME REVERSAL AND OTHER QUESTIONS

174. Two related questions can be considered together:

Question 1. Is the impossibility of a time reversal of events still explained by probabilistic arguments?

Question 2. Consider for the sake of argument an isolated system involving macroscopic objects, that is initially in a pure state and must remain so according to the Schrödinger equation. When looking only at collective observables, we are confronted with a mixed state after decoherence. I can, however, always conceive of an observable whose measurement would reveal the persistence of the pure global state. Does not this obvious remark imply that decoherence is only an appearance and not a fundamental explanation of the Schrödinger cat paradox?

Answers. The two questions are related, but I will answer them successively. Let us begin by recalling the traditional explanation of the direction of time in statistical physics.

The difference between an initial and a final state in an irreversible process lies in the probabilities of going from one to the other. The initial state is produced through a control of its collective parameters, and it automatically leads after some time to a state showing the collective parameters of the final state. In classical physics, the initial state would necessarily reappear after an excessively long Poincaré recurrence time. This is only true, however, if the system remains isolated. It cannot be: Only the universe itself is isolated (?) during such a long time. The question thus becomes one in cosmology and I will not discuss eternal return as a serious problem in physics.

Another version of the same argument is well known: why not reverse all the molecules' velocities so the system will go back to its initial state? The answer is again probabilistic: among all possible velocity distributions, the ones returning to the initial state have a negligible weight. In other words, the probability of producing the desired time-reversed (velocity-reversed) state when one can only control some collective parameters is negligible. According to Borel, it must be considered as zero. These arguments are unchanged in the present interpretation, except that the previous probabilities are still incredibly much smaller when decoherence is taken into account. But, their exact value does not really matter, according to Borel.

Let me now turn to Question 2: untying decoherence. One can certainly conceive of an observable revealing the persistence of a global pure state. One can say such words as: "Suppose I measure this observable." But one cannot *make* the measurement. For any not-too-small system, one can mathematically construct various observables for which a yes-no measurement would give, in the case "yes," evidence for the persistence of the pure state. The probability for getting the "yes" answer is, however, extremely small and the probability of "no" practically 1. The conclusion is obvious.

There is another answer given by Asher Peres (1980) in a paper entitled "Can we undo quantum measurements?" He shows that an external apparatus restoring the pure state would not obey the second principle of thermodynamics.

The gist of the answer to every similar question is always the same: compute the probability for realizing the effect one is thinking of. There is every reason to believe that it will always be so small that Borel's axiom can be applied. There is no indisputable proof of this, however, and some clever person might find a counterexample: it occurred once with the spin echo effect where an apparently lost signal was recovered. A reasonable assumption is that any counterexample would be a new, nice physical effect but with no implications for the general case.

THE STATUS OF FACTS

175. A consequence of quantum dynamics, although less well known than the paradox of Schrödinger's cat, is certainly as worrying. We met an example in Section 162 where the state of an ideal detector was changed through a later action on a distant particle. This result seems to imply that anything having apparently happened can always be contradicted, suppressed, or erased, some time later as if it had never existed. A dead cat can become a living one. The outcome is that

quantum mechanics seems to exclude the existence of the most basic physical notion in empirical physics: the notion of facts.

The exact reasons for this rejection are contained in von Neumann's formulation of properties, measurements, and interactions. Consider a property of a system, for example, "that cat is dead," which is represented by a projector. Von Neumann, showing himself in that instance more a mathematician than a physicist, assumed that every observable can be measured and every Hamiltonian can be realized, at least in principle. If the second assumption holds, it is clear that one can always devise a Hamiltonian bringing back the system to satisfy some time later the property "the cat is alive."

For a cat having one degree of freedom, or equivalently a pointer, the action of a potential is enough for producing a see-saw motion between two positions, or a forward and backward change between "dead cat" and "alive cat." The double-well potential shown in Figure 20.1 can do that: starting from a position inside one well, the system will be found in the other well after some time. One can also act on a third party as we did in the previous chapter by bringing two separated wave functions for an atom into a common one. Whichever tricks one uses, this is always

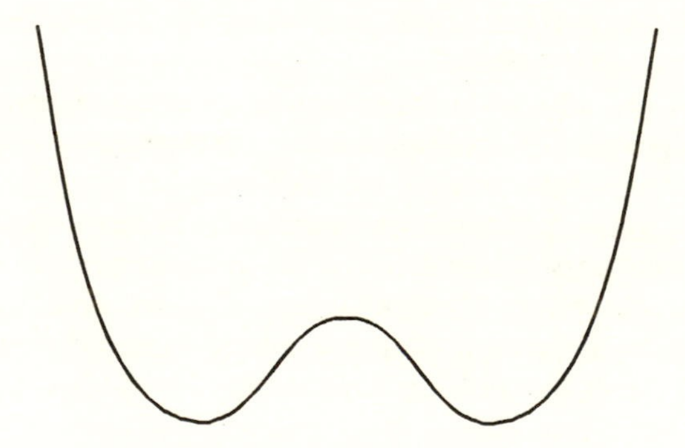

Figure 20.1. A double-well potential. A particle in a double-well potential can ideally represent a Schrödinger cat. When the particle is in the left-hand well, it stands for a dead cat, and it stands for a live cat when in the right-hand well. It can however cross the barrier and be found alive again after having looked dead. Such a non-factual behavior is impossible in a real measurement when there is decoherence as an effect of interaction with the environment.

possible. This is a really dramatic result since a conflict with the existence of facts is the worst possible inconsistency for a physical theory. Hence the following question:

Question. Can interpretation agree with the empirical existence of facts?

Answer. The answer was essentially given in the previous chapter but I will elaborate on it. It should first be stressed that the notion of fact does not belong properly to the domain of theory. A theory is a description. It belongs to the highest category of language, and as the saying goes, the word "dog" does not bite. What one must ask of a theory is that it provide a notion describing what we see as fact and that nothing in this theoretical notion should conflict with what we observe as facts.

Moreover, quantum theory like any theory or any language does not deal with existence but with possibilities. When it envisions a property, it can always envision its contrary as well. When a fact occurs, we must always consider that the language we use may propose two boxes on a piece of paper, one with the words "this fact occurs" and the other with "this fact does not occur." We mark one of them when the fact really occurs under our own eyes (may the reader forgive my insistence on such trivialities, which have often been found a source of confusion).

I will not try, therefore, to define a fact, because a primary notion can never be defined and there is nothing more primary in physics than the notion of fact. I can, however, define what we mean by a quantum property that can *describe* a fact. It is a classical property (as understood in Chapter 10) concerning some collective observables in a macroscopic object, this property being definitely separated from other exclusive properties of the same kind by the decoherence effect.

We saw in previous chapters the importance of each condition entering into this definition. On the other hand, it is obvious that the facts we see can be described in that way in a quantum framework. A further analysis of this notion requires a definite language, and I will use the one given by histories. I will call a property satisfying this definition a "factual property," insisting accordingly on its use in language. It does not preclude the consideration of other similar properties referring to the same value of time and describing other facts, exclusive of this one (a memory computer showing voltage zero rather than voltage V).

The main feature of decoherence is that it keeps exclusive facts separated. This feature is a result of the diagonal character of reduced states. We saw also that a later action cannot spoil this diagonality, at least if we accept Borel's axiom. A family of consistent histories is split once and for all into separated histories when a decoherence effect has occurred. This means that records (i.e., factual properties that are deterministically defined by a previous one) can exist. Even if every record has been erased by dissipation, no later factual property that would contradict the first one can satisfy logical consistency.

I conclude that the logical conditions imposed on our language are in full agreement with the observed qualities of real facts. Facts have a definite meaning in interpretation, the word "meaningful" being equivalent to "allowing logical consistency."

A last word of caution should be added. There are cases where decoherence is weak (i.e. for a rather small system after a rather short time). Physicists must exert much ingenuity to realize these conditions, but one must take them into consideration. The question is then: Can there be facts in that case? My answer is: there are no more facts in that case than if there were no decoherence at all. The properties one may consider have some finite probability p, and nothing can be asserted before the occurrence of a later fact (or the introduction of a later factual property), which has an a priori probability equal to p. Nothing is acquired until the probabilities of simultaneous existence of distinct factual properties (nondiagonal elements of a reduced state) are not small enough for escaping any possible measurement—when the Borel argument becomes applicable. Of course, one must define a "very small" probability for a quantum property. The most convenient semiempirical definition is to say that it is one that cannot be measured with any apparatus involving, for instance, no more matter as there is in the present universe.

OBJECTIFICATION

Stating the Problem

176. The problem of objectification can be stated as a question:

Question. Why is there a unique datum at the end of a measurement?

Answer. The word "datum" means a special kind of phenomenon: the one shown by a measuring device (which will be defined more properly in Chapter 21). Objectification is what remains of the reduction effect after the discovery of decoherence: There were two aspects in the reduction of a wave function. On one hand, it gave a rule for computing the probability of a second measurement result when a first one is granted. This we found to be a direct consequence of decoherence, as shown in Chapter 20. On the other hand, the rule also asserted the actual uniqueness of the first datum, and the origin of this uniqueness has become the problem of objectification.

When Bohr spoke of reduction as a rule with no analogue in science, he did not mean its statistical aspect but rather the resulting uniqueness of data. Some critics of the meaning of decoherence have retained this view, (e.g. d'Espagnat, 1994). According to them—and even if one agrees that decoherence justifies the statistical use of the reduction rule—the problem of objectification is not solved, and they consider it as more fundamental than decoherence. In that sense, quantum mechanics in the present interpretation would not be complete, and once again, the ghost of incompleteness reappears.

The question is important and I will later mention the various solutions that have been proposed. Before I do so, however, I will discuss the answer arising most easily from the present interpretation. It sounds rather drastic: there is no problem of objectification, because the relation between a theory and physical reality is no part of a theory. The conditions of this relation must be *added* to the theory itself, and this is where the requirement of uniqueness enters. Uniqueness must be prescribed and one cannot expect to find it directly in the theory. The only sensible requirement is that uniqueness be *compatible* with theory, and it is.

This sweeping answer will be enforced by showing also that the supposed problem of objectification cannot be stated in a logically significant way. It is not a problem of physics and only a problem of interpretation, which means that the analysis to which one must proceed is not a matter of theoretical physics but rather of semantics: a philosophical exercise, so to speak.

Rejecting the Problem

177. I have made strong statements and they must be argued carefully. I will therefore clearly distinguish various steps and aspects of the argument.

The GGH Argument. I will begin with some methodological aspects of the problem. One should certainly quote first an analysis by Griffiths (1984) and by Gell-Mann and Hartle (1991), which can be called the GGH argument. It starts from the recognition of quantum mechanics as a basically probabilistic theory. Then, one cannot ask from the theory to provide a cause, a mechanism, or an explanation for the actualization of events. Anything specifically actual does not belong properly to quantum theory. This is a strong argument, but it did not seem to make many converts, and I will have to push it farther before it becomes compelling.

The Power of Causality. When contemplating a fundamental question, one may sometimes wonder why it is considered as fundamental. This is a significant question in the case of objectification, which is often hailed as a touchstone of physics. Some people went as far as calling it *the* unique problem in interpretation, and one may wonder from where their conviction came.

One cannot attribute the quest for an objectification mechanism to an influence of classical determinism. Laplace's doctrine does not have this strength of persuasion. There is something deeper and everybody can feel in oneself a belief in the reality and importance of the problem: it comes from the working of our own minds.

There is in the attractiveness of objectification a strong desire for making it the result of a mechanism, as the concept of something obeying causality. This category of causality is so much part of our minds that any violation of it urges us to find an explanation. One cannot forget, however, Pauli's criticisms of Kant's categories and an interesting possibility arises: Could it be that the problem of objectification is logically unsound? This would be a stronger version of the GGH argument, and I will explore this avenue, which has apparently never been considered previously.

The Uniqueness of Reality Is Always an Assumption. Although remaining at a level of preliminaries, I will now consider another aspect of the GGH argument. It has to do with the relation between theory and reality. I will speak presently of "physical reality" as meaning the world of phenomena (which can be directly perceived by definition) so that physical reality (Other people say "empirical reality") is necessarily macroscopic. Our access to this world is purely empirical.

A theory relies on a vast collection of observational data. It tries to account for the correlations among the data in terms of basic laws. When a physical theory is more than a collection of empirical rules, it provides a correspondence between physical reality and mathematics. (One may notice, by the way, that the existence of this kind of correspondence remains extremely mysterious in itself, but this is not my topic.) The theory is then a mathematical representation of the phenomena, and it can also involve other necessary notions—such as atoms, particles, and fields—which may not be directly empirical.

The uniqueness of physical reality is never a consequence of a theory, rather this uniqueness must be *added* to the theory. This is a very important feature of knowledge, which is true even of pure classical physics. Classical physics is perfectly deterministic, but this does not imply by itself the uniqueness of reality. One must add that at some initial time, the state of a system is unique, that is, physical reality is unique. This assumption is not however compelling since one could very well conceive of a physics rejecting it; for instance, the framework would be a synthesis of classical dynamics with a probabilism originating in chaos, a view that is perfectly sensible and consistent. The distance between theory and reality is more obvious in the case of quantum theory, and the connection between theory and reality must be investigated as a specific aspect of physics, that does not belong to theory alone.

The "Category" of Reality. Let us then consider this relation more closely. Because theory relies on general laws, it covers in principle any conceivable collection of phenomena in a definite field. Conversely, what can be expressed through the language of the theory is not restricted to actual events but also extends to potential ones. This language can envision various possibilities, events which could exist, which have existed or will exist, or simply will never exist although they do not contradict the laws.

With its specific language, the case of quantum mechanics is exemplary. We saw that this language can account for any correlation between real data, but it has a wider range. As a language (and in fact as any language), it deals with possibilities. Better than most languages, it does not allow nonsense, and finally, like any sound language, it can express the assertion of a real phenomenon by a special kind of statements.

It is obvious that a proposition directly related with physical reality is special. This peculiarity is precisely the content of Kant's *category* of reality: One might say that some sentences can be written in red ink to emphasize that they assert something as being real and not only possible. These propositions are also said to be *true* and they express *facts*.

Two Questions. Before investigating how physical reality can be asserted as unique, one may wonder whether this is necessarily true. Perhaps reality is not unique after all and quantum theory cannot decide. The example of Everett's many-worlds conjecture (which will be described later) leaves this possibility open, at least in principle. When stating the problem of objectification, we must therefore consider as distinct two questions: (1) Can one can make explicit the category of reality (i.e., the use of the red ink pencil)? (2) How can one assert furthermore the uniqueness of reality?

The Category of Reality Is Valid. Let us begin with Question 1 as seen in the framework of the present interpretation. The answer is mainly a summary of a work we did already. We saw how classical physics can be derived from quantum mechanics and have also seen how its proper common sense logic can be obtained. It was said in Chapter 16 that this amounts to recovering the category of reality and that there is no possibility of internal contradiction if one uses assertions of the type: "this definite classical property expresses a real phenomenon."

As long as one is considering a macroscopic situation, where the quantum-classical correspondence theorems hold and no quantum measurement happens, it is possible to add "this is real" for the relevant properties describing physical reality.

Macroscopic Reality Can Be Assumed Unique. Let us now come to Question 2, that is, to the question of uniqueness. I will consider first the previous situation where everything macroscopic behaves classically and there is no measurement. We saw in Chapter 16 that the language of interpretation is *compatible* with the uniqueness of physical reality, but we must now take into account that this does not mean that interpretation implies or even necessarily requires this kind of uniqueness.

Our language can easily envision logical alternatives concerning conjectures. There is, for instance, no problem in expressing by a classically meaningful projector E_1 that the Empire State Building location is in New York. Another projector E_2 can state that the building is in San Francisco. The sentence saying that "the Empire State Building is in New York or San Francisco" makes sense, and a child ignoring the reality of geography could say it without violating logic or grammar. This is represented by the projector $E_1 + E_2$. But when one comes to the real Empire State Building, no theory can decide where it is. This is a purely empirical matter and its uniqueness must be asserted independently from the rest of the theory.

There was no principle concerning empirical realty among the axioms of quantum mechanics; we were then in a world of theory. The relation between theory and reality, the assumption of a correspondence between them, is a principle of another order, a purely empirical statement whose main feature can be asserted by a

Fundamental Empirical Rule. *The phenomena we observe are unique.*

This new principle can be supposed valid for ordinary phenomena (when there is no measurement). It does not follow from quantum theory, but it has been proved to be compatible with interpretation, and this is essential.

Uniqueness Holds Also in a Measurement. We now arrive at the standard problem of objectification, which is concerned with the outcome of a quantum measurement. We saw in Chapter 5 why the Copenhagen tradition kept classical phenomena and quantum properties separate. Two different languages described them: an ordinary classical language and another relying on the complementarity principle. One could not then avoid the objectification problem, which was "solved" by state reduction. But we now dispose of a universal language for interpretation, and the situation becomes different.

Let us consider an example where a quantum measurement can yield two different results, each of them giving a datum with, for instance, two different positions, 1 and 2, for a pointer on a dial. Suppose that in position 2, the pointer can touch an electric contact, the event resulting in an electric current which detonates a bomb that destroys the Empire State Building.

This is, of course, a situation similar to the one with our old pet, the Schrödinger cat. We know that an essential consequence of decoherence is that the two possibilities "the building is safe" and "the building is destroyed" are genuine alternatives. If we agree on the fundamental nature of decoherence, we are then back exactly to the question we met previously when there was no measurement. There is no logical difference for a building to be in San Francisco rather than in New York or to be destroyed rather than undamaged. In both cases one must complete the theory by empirical input to decide whether or not physical reality is unique.

The argument is the same as in Section 175: Different measurement data exclude each other (their projectors are mutually exclusive). Any one of them cannot be negated by later events (because of decoherence and the Borel axiom). One can thus extend the range of the fundamental empirical rule to the results of a quantum measurement. There is nothing new, nothing more. The rule is perfectly compatible with observation and compatible as well with the theory and its language. Objectification is not a problem of theoretical physics; it states an empirical feature of reality that must be added to the axioms of the theory for relating it to actual experiments.

Another Approach. I will conclude by reconsidering the problem from a logical standpoint to show that the statement of the objectification problem does not make sense in any language we know. This argument is independent from the previous one.

The problem is usually stated in a standard language, namely, English in the present case. But we know that plain English can express only macroscopic physical reality; it does not properly describe atoms or particles because of complementarity. The importance of this difference becomes obvious when one states the objectification problem as in the following question: after an interaction of the electron with a measuring device, how can a measurement giving the position of an electron give a unique datum?

The word "electron," and particularly its position, do not belong to plain English because they are exposed to the ambiguities of complementarity; however, they are necessary to make clear what the datum is about. The word "interaction" is clear only when referred to its mathematical expression, and it is also needed to make clear what is meant by a measurement. The sentence stating the problem of objectification, when it is stated completely, does not therefore make sense because it combines several different languages, two of them (about the electron position and about the datum) being from different categories.

When one tries to put the two languages together, one obtains the universal language already considered at length. The description of a measurement then makes sense, as Chapter 21 will show, but we have just seen that its way of answering the problem of objectification is trivial and can be replaced by a simple requirement of compatibility. There is no language, as far as I know, having a sound logical basis in which the question of objectification might be stated properly, so we cannot even say that the problem makes sense.

This answer should not be too surprising. This is not the first time when an apparently obvious or cogent question is found to be nonsense. One may remember on this occasion the question of what happens "during" the emission of a photon by an atom or "through which hole" a photon is going. These questions looked perfectly natural and the people who asked them were perfectly sensible. The answer from quantum mechanics was that in spite of the habits of our mind, the questions do not make sense. It seems as if the same conclusion would hold for objectification.

OTHER VIEWS ON OBJECTIFICATION

178. One of the entertaining aspects of interpretation is its lively controversies. The question of objectification is particularly acute, and I will now review briefly the main approaches that are opposed to the one I just proposed. There are essentially four main approaches: (1) Everett's approach, assuming a multiplicity of realities. (2) Theories assuming a realistic reduction process. (3) Hidden-variable theories, in which reality is automatically unique. (4) The point of view indicated above, namely that objectification is a matter of compatibility and not of necessity.

Everett (1956) pushed the superposition principle to its ultimate consequences. He assumed a wave function of the universe and a strict Schrödinger equation in which every part of the wave function has a real counterpart. He therefore considered the universe as being multiple: there are many different copies of the universe, some of them with different copies of ourselves. They separated with different branches of the universe wave function when some event in the past acted as a quantum measurement. Physical reality became multiple. The decoherence effect (of which Everett was a precursor), however, forbids every kind of communication, of interaction between the various branches of reality, which represent the components of a gigantic and unique wave function. The main virtue, as well as the defect of this approach, is that no conceivable experiment could ever prove it to be right or wrong. It looks like a pure Platonic Idea (since one may remember that Plato considered ideas, or in some sense theories, as more real than the empirical world we see). The Idea, the absolutely real thing is now called *psi*; it is the wave function in which every possible knowledge about every possible universe is contained. There is of course a strong metaphysical flavor to this vision.

Let us now consider the theories that assume a physical objectification process. G. C. Ghirardi, A. Rimini, and T. Weber (1986 GRW) assumed that such a process is acting randomly and can happen at any time anywhere. Of course it does not obey quantum mechanics. The effect acts in practice like the reduction process did in the Copenhagen interpretation, although it is now random and not especially associated with a measurement. It changes the wave function of any particle in a random and sudden way. More precisely, this wave function becomes zero at a random time outside a small random space region with a typical size a and, inside this region, the wave function is essentially unchanged, except for a multiplication by a factor preserving normalization. This realistic reduction effect is supposed to occur with an average time interval T between two reductions. The average parameters a and T must be cleverly chosen so as to preserve the validity of quantum mechanics at the atomic scale and to avoid large quantum effects at a large scale. The proponents of the theory claim that this fine tuning can be achieved, and as far as I know, there is no compelling counterexample.

Other proponents of a realistic reduction process, or R process, insist on the objectification of space-time. According to Einstein's equation for general relativity, a quantum superposition of several states of matter representing different macroscopic locations should imply a superposition of different space-times. The R process is supposed to avoid this secondary superposition, and because of its link with general relativity, it is dominated by gravitation. This approach has been developed particularly by F. Karolyhazy (1966, 1974, 1986), Phillip Pearle (1976, 1984, 1989), and Roger Penrose (1996, 1997). It is very interesting because it brings forth a new aspect of the objectification problem with its relation to space-time geometry. I will return to it in the next section.

David Bohm (1966, 1993) investigated and developed another approach that had been initially proposed by de Broglie. He assumed every particle to have a definite position in space and a definite velocity. This is intended explicitly to make the properties of the particles plainly real. Bohm accepted, however, the notion of a wave function Ψ obeying Schrödinger's dynamics. The real particles themselves obey Newton's dynamics, in which the wave function contributes by creating a quantum potential into which Planck's constant and the wave function Ψ enter. The peculiar feature of the theory is that this action of the wave on the particles has no reaction and that the real position of the particles has no influence on the wave function. When thinking of charts showing data in various colors (the ground temperature in meteorology, for instance), one might say that according to this theory the wave function is colored by particle trajectories. I have already mentioned the main difficulties of Bohm's approach. It does not provide a satisfactory field theory because one must decide beforehand what is real, either the photons or the electromagnetic field for instance. The treatment of spin is also awkward, except if one puts spin directly in the wave function. David Dürr, Sheldon Goldstein, and N. Zanghi (1992a, b) have, however, shown that the "real" motion of the particles is necessarily so complex that their position must be described by probabilities. These probabilities, furthermore, coincide with the quantum ones, as obtained from the wave function.

COULD THERE BE A REAL OBJECTIFICATION EFFECT?

179. Although there is much personal bias in the interest I have for the various theories I briefly described, the ones assuming a realistic R process for the uniqueness of space-time seem to point out the most cogent physical questions. Of course, it may be argued that the uniqueness of reality, which agrees with consistent histories, would also agree with the uniqueness of space-time. The proposal I made earlier also answers, therefore, Penrose's questions. I think nevertheless that the ideas introduced by Karolyhazy, Pearle, and Penrose are extremely interesting, and I shall therefore discuss them more carefully in this section.

The argument consists in an analysis of the conditions to be expected for an R process. It involves three steps:

1. The typical time of action of the R process must be of the same order as the decoherence time.
2. The R process must be triggered by decoherence.
3. The explicit action of the R process is essentially unique.

The conclusion will be the existence of very strong constraints on the presumed R process, making its existence apparently impossible. The present argument should be considered as a critique rather than as a criticism. This means that if somebody were able to satisfy the constraints we find, the present analysis could become a step in the construction of a new theory of objectification.

The Time of Action of the R Process

We have seen that decoherence is extremely efficient in cutting off the superposition of different phenomena. The decoherence effect exists; it has been directly observed and it cannot be ignored. It is certainly by far the most rapid macroscopic effect and so raises the question of comparing its duration with the time necessary for an assumed R process. One may notice that the proposals that have been made for the R process make it much slower than decoherence. To give an example, the photographic detection of light amounts to the displacement of some millions (a small number!) of silver ions aggregating along a defect in an emulsion grain. The ions turn into silver atoms after fixing electrons. The theory of decoherence tells us that quantum superpositions initially present in light have been destroyed as soon as a few atoms have formed (and even when only one ion has become an atom, at the extreme). This is a very short time. On the other hand, the Ghirardi-Rimini-Weber (GRW) or the gravitation processes would take weeks before objectification. We know in any case that the R process is certainly not quicker than decoherence. This is a consequence of the Haroche-Raimond experiment (Brune, et al. 1996, see Chapter 22) because any process really competing with decoherence would spoil the excellent agreement between the experimental results and decoherence theory.

I must now introduce a principle of economy—Okham's razor. It seems very uneconomical to have two processes doing the same job. If the typical times for decoherence and for the R process were substantially different, we would have the following situation: Decoherence acts first and suppresses every experimental consequence of a quantum superposition, leaving the remains of superposition as exponentially small nondiagonal terms in the collective state. Then, somewhat later, the R process would come and replace a result that is valid for all practical purposes by a perfected one, now valid for all philosophical purposes. The only sensible assumption is that the two processes have a similar characteristic time.

Decoherence as a Trigger

The characteristic time of decoherence depends strongly on the detailed coupling between the collective subsystem and the environment. In some cases, it depends on collisions; in other cases, on electromagnetic effects; sometimes on losses outside a reservoir; and so on. This versatility can only be present in the R process if its action is controlled by decoherence. In other words, decoherence must trigger the R process.

Everybody agrees that objectification would violate quantum mechanics through corrections to Schrödinger's dynamics. If, by some mechanism we don't understand, decoherence amplifies a dormant process, the R process could be very weak in ordinary quantum conditions, whereas an amplification through decoherence could make it very large.

Gravitation effects are then good candidates for entering into the R process. The reason is that they are sensitive to space densities, and we have seen that decoherence easily accommodates spatial collective variables. We also saw that this accommodation is ultimately related to the invariance under a change of reference frame, which is essentially the principle of inertia with its gravitational flavor. Furthermore, position variables are the only collective ones we know to be universal. We might even assume that any kind of decoherence is coupled to a specific decoherence in position observables (the position of a group of silver atoms in a photograph, for instance).

What would then be the mathematical expression of the triggering effect? Many quantities vanish exponentially when decoherence is almost complete, for instance, a trace such as $\mathrm{Tr}(\|[\rho_c, Y]\|)$ (notice the absolute value) where Y depends on Newtonian positions (it can be the gravitational energy). Many of these quantities vanish exponentially with time, and if one of them is part of a denominator in the mathematical expression of the R process, this process will be triggered very efficiently.

There is however a difficulty. For all the triggering functions one can think of, their disappearance occurs when one uses the reduced state operator ρ_c, but not the full state operator ρ. Remember that $\rho_c = \mathrm{Tr}_e(\rho)$. But the partial trace on the environment is a linear operation with very constraining algebraic properties. We cannot envision that the R process, which is supposed to be fundamental, can distinguish between what is a relevant *macroscopic* position and what is the environment. One is therefore confronted with a serious *algebraic difficulty* when trying to construct a theory of the R process: there is no known quantity involving the full density operator, which would vanish with decoherence (or become infinite) and could be part of a triggering mechanism.

The Effect of the R Process

To see what the R process does, it will be convenient to consider an example. Consider a measurement process of which the end result can be one of three possible positions $(1,2,3)$ of a pointer on a dial. I will disregard the environment (which would lead us into technicalities) and concentrate on the reduced density operator. After decoherence, the state is practically a 3×3 diagonal matrix with diagonal elements (p_1, p_2, p_3), which are the respective quantum probabilities for the three possible results. Suppose now that the R process acts, and in a specific instance, the pointer ends up in position 1. Quantum mechanics dominates again, and subsequent events must start in that case from the diagonal density matrix with elements $(1,0,0)$. There is however a very strong constraint on the R process: When repeated many times under identical conditions, it must act randomly (since we see random events), and its probability for ending with the matrix $(1,0,0)$ must have the quantum value p_1. Otherwise, quantum mechanics would not agree with observation.

Except for this constraint, namely randomness with definite final probabilities, we may assume the R process to be continuous in time, which means that it is a dynamical process. I will now consider two questions. The first one is the origin of randomness, and the second one is a more precise prescription for the process itself.

The GRW process, for instance, is classically random. It would be awkward though, to see an intrinsic classical randomness enter into the answer to a supposed difficulty of quantum mechanics, whose origin lies in the intrinsic probabilism of the theory. There are, however, many sources of random quantities, for instance, in the inaccessible properties of the environment, and I will therefore assume that something like this is at work for generating randomness. To say what it is exactly is a difficult problem, but its existence is not in itself a fundamental difficulty.

The R Process as a Random Walk

I wish now to argue that the dynamics of the R process is strongly constrained: it must be a very specific kind of Brownian motion. Being random and continuous, it can always be assimilated with a Brownian motion whose law (the correlation functions for infinitesimal changes) may vary with time and with the moving parameters (i.e., p_1, p_2, p_3 in the present case).

The argument will be clearer if the kind of motion I have in mind is proposed first as a particular model. This is a version of a model that was first introduced by Pearle (1976, 1984), and it is best explained in geometrical terms. I refer to the previous transition from the diagonal state (p_1, p_2, p_3) to the reduced states $(1,0,0)$, $(0,1,0)$ or $(0,0,1)$. It will be convenient to draw an equilateral triangle with height 1, the three corners being denoted by $(1,2,3)$. The initial diagonal density can be represented by a point P inside the triangle whose distance to the side $(2,3)$ is p_1, with similar values for the distance to the two other sides. Assume now that point P moves with an isotropic random walk in the triangle (by isotropic, I mean with respect to the Euclidean metric for the plane of the triangle). It must necessarily hit one of the sides of the triangle after some while, say the side $(1,2)$ for definiteness. Then P has a random walk on this side, until it reaches a corner, say point 1, for instance. The beauty of this model is that the probability for point P to finally reach the corner 1 is exactly p_1. It goes without saying that it then represents the density matrix $(1,0,0)$.

I will sketchily indicate an instructive proof of the result, which will be useful for the discussion. Let P_o be the starting position of the moving point P. Let Q be a point on the triangle boundary. Let $df(Q, P_o)$ be the probability for P to hit the boundary (necessarily for the first time) in a definite infinitesimal interval of the boundary containing Q. Let C be a circle with center P_o in the interior of the triangle. Let M be an arbitrary point of C. We denote by $g(M)\,dl$ the probability for the moving point to hit the circle for the first time in an infinitesimal interval of length dl centered at M. Combining probabilities, one has obviously

$$df(Q, P_o) = \int_C g(M)\, dl\, df(Q, M) \Big/ \int_C g(M)\, dl.$$

But since the Brownian motion is isotropic, g(M) must be a constant so that

$$df(Q, P_o) = L^{-1} \int_C dl\, df(Q, M), \tag{20.1}$$

where L is the length of the circle. It is well known that equation (20.1) is a necessary and sufficient condition for the function $df(Q, P_o)$ to be a harmonic function of P_o (i.e., its Laplacian vanishes).

This is the essential point of the proof. One obtains the final result, that is, probability for ultimately reaching one of the three corners, by an induction on the number of dimensions. In dimension 1 (when the triangle is replaced by an interval with length 1, which we may take to be the interval [0, 1]), the infinitesimal df becomes finite and one writes, for instance, $F_1(P_o)$ for the probability of P to reach the extremity 1 of the interval when starting from P_o. Being a harmonic function, it is given by $ap_1 + b$, where a and b are constants. Since $F_1 = 0$ for $p_1 = 0$ (or $F = 1$ for $p_1 = 1$) when the point starts from one of the two ends, one must have $F_1 = p_1$. This result can then be used as a boundary condition for the case of the triangle, from which one goes to a tetrahedron and so on.

Onc might ask what the purpose of this example is. It consists in inverting it and looking at its assumptions as being the result we must get: we know that the probability for (p_1, p_2, p_3) to reach $(1,0,0)$ must be p_1. But p_1 is a harmonic function, so that the random motion must satisfy equation (20.1). This can be shown to imply (nontrivially) that the motion is isotropic, the only freedom we have being its time rate. It is fortunate that the result does not depend on the rate, because the triggering of the R process is presumably changing rapidly with time.

This result shows that the motion must be isotropic, so that it is essentially unique. The idea of an isotropic motion is in fact rather natural. It has an operator meaning since it corresponds to a metric on the space of density matrices given by $\mathrm{Tr}(\delta\rho^2)$, if $\delta\rho$ is an infinitesimal change in the density. One should also not forget that decoherence can still act during the R process, so that it maintains diagonality. Our assumption about point P staying on the triangle boundary after reaching it is also a consequence of the theory of decoherence: it never takes the system out of the sub-Hilbert space in which ρ_c is nonzero (this property can be shown by using equation (17.6), but I will not give the proof).

Summarizing the Results

There are thus very strong arguments for the R process to be triggered by decoherence and to have an essentially unique behavior. The existence of a sensible (i.e., universal) mathematical expression for this process is on the other hand very questionable. An algebraic difficulty was already mentioned. There is a similar one with the isotropic random motion. It is very easy to write out explicitly in terms of the reduced density operator, but once again, we run into the difficulty of expressing it in terms of the full density operator. The reason is that the correlation functions for Brownian motion (or a more general stochastic motion) are quadratic in the coordinates (or velocities) of the moving point and so conflict with the linear character of the environment trace. Of course there would be further constraints when imposing general covariance, but we need not enter into them here.

When considering what little progress has been made in the production of a sensible R process without the constraints we have exhibited, its existence appears as extremely questionable. Resorting to quantized gravity or something of that sort as a way out does not seem of much avail because the physics we are talking about is here, in the laboratory, and it works at a large scale. So, in conclusion, one may consider the existence of a physical reduction process as extremely doubtful.

More generally, the choice between the existence of an objective R process, or the Everett answer, or the treatment of objectification we found earlier is a choice between a reality of the wave function (in the first two alternatives) or an empirical standpoint for which only the agreement with observed data and logical consistency are only required. There is therefore little doubt that the controversy will stand for a long while.

SUMMARY

After the observation and the theory of decoherence, most so-called fundamental questions obscuring a full understanding of quantum mechanics depend on the epistemological status of extremely small theoretical probabilities. An interpretation of probability calculus must therefore stand at the entry to an interpretation of quantum mechanics. The most convenient one was proposed by Émile Borel: an event with too small a probability should be considered as never occurring. From an empirical standpoint, a very small probability is one that cannot be measured by any experimental device which can be realized, or even conceived of in the universe.

One can then assert that data separated by decoherence are separated forever, except for very small systems. A definite universal direction of time therefore exists at a macroscopic level. It is the same for decoherence, dissipation, and the logic of quantum mechanics. The no-return character of decoherence implies that phenomena cannot be contradicted by any later action or event. They accordingly have the character of facts.

The objectification problem (i.e., why there is a unique output datum in a measurement) can be considered as a false problem. It arises from a traditional line of thought with no compelling logical foundation. The universal language of interpretation can easily envision a unique physical reality (of the phenomena), in agreement with observation.

21

Measurements

This chapter concludes interpretation with its most important application: a theory about measurements allowing a direct exploitation of experimental data. The method we used, being demonstrative at every step, converts the main results into so many theorems. They are essentially the same as in the Copenhagen interpretation, where they were considered as independent principles. One can further notice three main differences:

1. The logical equivalence between an empirical datum, which is a macroscopic phenomenon, and the result of a measurement, which is a quantum property, becomes clearer in the new approach, whereas it remained mostly tacit and questionable in the Copenhagen formulation.
2. There are two apparently distinct notions of probability in the new approach. One is abstract and directed toward logic, whereas the other is empirical and expresses the randomness of measurements. We need to understand their relation and why they coincide with the empirical notion entering into the Copenhagen rules.
3. The main difference lies in the meaning of the reduction rule for "wave packet collapse." In the new approach, the rule is valid but no specific effect on the measured object can be held responsible for it. Decoherence in the measuring device is enough.

The most important results were already obtained in Chapter 19, and the present chapter will be essentially a summary providing some practical rules and a few comments. No proofs or calculations will be shown but some formal aspects, which may raise a question or give more precision, are mentioned in a smaller type.

A MEASUREMENT EXPERIMENT

180. I will begin with some conventions and notations.

There are many different kinds of experiments in quantum physics, but we are interested only in the ones ending with the empirical evaluation of a probability (whether every experiment is of that kind in some sense or not is of no concern to us). Such an experiment requires a large number of individual measurements for collecting statistically significant amounts of data. I will proceed as if each measurement were made on a unique atom or a unique particle and is not an average of many of them. The microscopic object undergoing measurement is denoted by Q.

The measurement is intended for obtaining the value of some observable A belonging to Q. I will primarily discuss the case where it has only discrete eigenvalues a_n. The measurement device is assumed to be perfectly efficient, always giving a datum when interacting with a Q system. This is, of course, an ideal situation, and in practice one must be aware of a possible lack of efficiency when discussing systematic errors. The measuring device is denoted by M.

It is often convenient to distinguish three categories of measurements, each category including the previous ones:

1. Ideal measurements: When the measured system enters into the measuring device in an eigenstate of A, it leaves in the same state.
2. Conservative measurements: One assumes that the measured system is not lost or destroyed during the measurement.
3. General measurements.

The final result of a measurement is a property stating: "the value of A is a_n." It is associated with a projector $E_Q^{(n)}$. One can treat similarly the case of a continuous spectrum by cutting it up into separate intervals, each of which is associated with a projector also denoted by $E_Q^{(n)}$.

Initial Conditions

181. We might try to retain full generality by considering a complete experiment with its preparing, confining, and measuring devices, but it would only complicate the discussion. I therefore simply assume that the measured system Q is in an initial state denoted by ρ_Q (or simply ρ when there is no risk of confusion) at some initial time, which is conveniently taken as the time when the measurement begins.

The measuring device is described by classical physics. It can show a datum after measurement (through a voltage in a memory, a pointer on a dial, a photograph, and so on). The various possible data are clearly separated from each other, so that the corresponding classical properties are clearly exclusive, as discussed in Section 92. A very important assumption is that decoherence has been fully active in the measuring device before the datum can exist (this of course implies irreversibility). As a matter of fact, no observer needs be mentioned.

When mentioning formal aspects, the notation ρ is reserved for the full state of $Q+M$. One may assume no correlation between Q and M so that $\rho = \rho_Q \otimes \rho_M$. There is a neutral property indicating that the apparatus is ready to work at the beginning of the measurement. This is a classical property (e.g., the pointer shows the indication 0 on a dial) for which one can introduce, as in Chapter 10, a quantum projector $E_M^{(o)}$. The observation of the initial neutral situation can be expressed by the relation

$$E_M^{(o)}\rho_M = \rho_M. \tag{21.1}$$

The various possible data are similarly represented by classical properties (n) and associated projectors $E_M^{(n)}$. Their mutually exclusive character can be expressed by equation (10.13) or by using Borel's axiom, allowing one to consider as zero a small enough probability:

$$E_M^{(n)}E_M^{(n')} = \delta_{nn'}E_M^{(n)}. \tag{21.2}$$

Moreover what we saw about classical physics implies that one can use without special care anything classical and deterministic in the working of the apparatus.

The Measuring Device

182. A measurement is a very special kind of interaction between the two quantum systems Q and M. Its essential character is that if the measured system enters into the interaction in a state which is an eigenstate n of A with eigenvalue a_n, the corresponding datum n is shown by the device after measurement. I should stress that this is a mathematical definition: the eigenstates one is talking about are mathematical assumptions, and the correlation between them and final data is a peculiarity of the Schrödinger equation for the whole system $Q+M$. Expertise in building and using instruments is, of course, the practical guarantee for this kind of condition.

This correlation can be expressed as follows: Let $|\psi\rangle$ be a mathematical initial pure state of the system $Q+M$ such that, at a time t_i when the measurement begins, one has the initial conditions $A|\psi\rangle = a_n|\psi\rangle$ and $E_M^{(o)}|\psi\rangle = |\psi\rangle$. If the time when the measurement is done is denoted by t_f, the condition for having a measuring device is

$$E_M^{(n')}U(t_f, t_i)|\psi\rangle = \delta_{nn'}U(t_f, t_i)|\psi\rangle. \tag{21.3}$$

PRACTICAL RULES

183. The description of the experiment is most conveniently done by using histories. This topic has been considered previously in some detail, and we can now concentrate on the measurements themselves. The histories must take the initial state (the preparation) into account, and they must introduce the various possible data, as seen at a time t_f when they are accessible. This time is controlled by decoherence. It is also convenient to consider what will be considered as the result of the measurement, namely, a property of the measured object Q.

The difference between datum and result is significant: A datum is a phenomenon, which we observe or can observe directly. It is a macroscopic property and is best expressed in classical terms, such as "the detector has registered." A "result," in the sense I use here, is a quantum property such as "the z-component of the atom spin is $+1/2$," something we cannot directly establish with our own senses and is the outcome of elaborate physical devices. When seen as time-dependent properties, the two notions are also different. The data do not exist before time t_f, which we have seen to result from decoherence. As for the results, it is convenient to consider them at a time t_i, when the interaction between Q and M begins.

One might wonder when exactly time t_i occurs. It is not clearly defined when, for instance, a somewhat diffuse wave packet controls the beginning of the interaction between an atom and an apparatus. But it does not really matter because we saw previously that there is some amount of freedom in the times entering into a history: it is a question of consistency. Nobody cares in practice because decoherence is such a rapid process; one can thus use in practice $t_i = t_f$.

The histories to be considered must contain at least the data properties with projectors $E_M^{(n)}(t_f)$ and the results expressed by $E_Q^{(n)}(t_i)$. Every history is described by an operator C_a and the family of histories is complete, that is,

$$\sum_a C_a \rho = \rho. \tag{21.4}$$

The consistency conditions

$$\mathrm{Tr}(C_a \rho C_b^\dagger) = 0, \qquad \text{for } a \neq b \tag{21.5}$$

are easily verified as a consequence of the measurement characteristic relation (21.3). Decoherence ensures the stability of the conclusions against later events.

Most of the analysis has already been discussed in Chapter 19, so the main conclusions can be stated as so many theorems in measurement theory.

Rule 1. *In a general measurement, the result is logically equivalent to the datum.*

This equivalence is of the type "datum $\Rightarrow$ result" together with "result $\Rightarrow$ datum" as it can be obtained in quantum logic from a consideration of conditional probabilities. Rule 1 justifies the usual assimilation of these two rather different properties.

The second rule is only concerned with an ideal measurement.

Rule 2. *If an ideal measurement is immediately repeated, it gives the same result.*

The third rule is Born's probability rule.

Rule 3. *The probability of the datum (or the result) n is given by*

$$p_n = \mathrm{Tr}_Q\big(\rho_Q E_Q^{(n)}(t_i)\big). \tag{21.6}$$

TWO SUCCESSIVE MEASUREMENTS

184. The case of two successive measurements on the same quantum object Q has already been discussed in Chapters 19 and 20. Briefly, those remarks can be summarized as follows: There is no physical effect on the measured system Q that might be called a reduction. On the contrary, the decoherence effect in the measuring device M is essential for a measurement.

There are two aspects of the theory of successive measurements. The first one is to compute the probabilities for the various results of the second measurement when the result of the first measurement is given. It requires only the computation of a conditional probability, which is easily obtained from the consideration of relevant histories taking decoherence into account. A second aspect, which is considered by some people as "the" problem of interpretation, objectification (the uniqueness of the final datum in the first measurement) was discussed in Chapter 20 and discarded on pragmatic grounds.

When these essential preliminaries are granted, the practical aspects of the reduction rule are rather easily expressed by

Rule 4. *Let a quantum system Q come out of a conservative measurement of an observable A. When computing the probabilities for the results of a second measurement on Q, one can apply Born's rule by using a well-defined expression for the state of Q at the end of the first measurement.*

To be precise, let t_f be the time when the first datum becomes available and t'_i the time at which the second measurement begins. Let us assume that an observable A has been measured first, then an observable A'. A specific result of the second measurement would be expressed by a projector $E'^{(m)}$ (there is no index reminding us which system is to be considered because it is always Q in the present case). Born's rule would state that there exists a state operator $\rho(t_i')$ for Q at the time t_i' when the second measurement begins such that the probability of the result m in the second measurement is given by $p_m = \mathrm{Tr}(\rho(t_i')E'^{(m)})$. Then to say that the state of Q is given at the end of the first measurement by $\rho(t_f)$ means that $\rho(t_i') = U(t_i', t_f)\rho(t_f)U^\dagger(t_i', t_f)$. The operator $U(t_i', t_f)$ represents the evolution of the system Q between the two measurements, under the external constraints to which it can be submitted.

The rule must be completed by the expression of $\rho(t_f)$. It will be convenient to proceed with increasingly more complex cases:

- When the result of the first measurement is a nondegenerate eigenvalue a_n of the observable A, one has simply $\rho(t_f) = |a_n\rangle\langle a_n|$ so that the probability we are interested in becomes $\|E'^{(m)}U(t_i', t_f)|a_n\rangle\|^2$. When the second result is a nondegenerate eigenvalue a'_m of A' and when either A or A' is a constant of motion, one obtains $p_m = |\langle a'_m|a_n\rangle|^2$.
- When the first measurement is perfect although the result is not a nondegenerate eigenvalue of A, one can use the Lüders formula (ignoring the short time necessary for decoherence)

$$\rho(t_f) = \frac{E^{(n)}\rho(t_i)E^{(n)}}{\mathrm{Tr}\{E^{(n)}\rho(t_i)E^{(n)}\}} \tag{21.7}$$

  The nonlinear occurrence of the initial state operator in the denominator must be attributed to the fact that one is dealing with a conditional probability in which the probability of the first event $\mathrm{Tr}\{E^{(n)}\rho(t_i)E^{(n)}\}$ occurs in a denominator.
- In the case of a nonperfect previous measurement, the rule is of little use. An example of a nonperfect conservative measurement is a charged particle crossing a detector (e.g., a wire chamber) and exiting it. It leaves enough track for its initial momentum to be known, but its momentum when leaving the detector is, of course, different. In such a case, the outgoing state one must use involves the first measuring device M as well as the measured system Q, and the state of Q involves the full $Q+M$ density operator. The relevant formula is:

$$\rho_Q(t_f) = \frac{\mathrm{Tr}_M\left(E_M^{(n)}\rho_{Q+M}(t_f)E_M^{(n)}\right)}{\mathrm{Tr}_{M+Q}\left(E_M^{(n)}\rho_{Q+M}(t_f)E_M^{(n)}\right)}. \tag{21.8}$$

The meaning of this formula is obvious: the state of Q at time t_f is obtained by tracing out the state of $Q+M$ in which the effect of the interaction between Q and M is taken into account. This is sometimes implicitly used in practice through classical considerations, for instance, when a heavy particle is first measured, then slows down before the second measurement. Once again, interpretation confirms standard practice.

LOGICAL AND STATISTICAL PROBABILITIES

185. I must now come back to an important point. The probabilities we have been dealing with up to now were introduced only for the purpose of logic. As a matter of fact, nowhere have we yet used the fact that the results occurring in a series of individual measurements behave randomly. This point was only briefly mentioned in Section 106, and it must be now considered carefully. It might seem as if the theory is using two different notions of probability—one with a logical purpose and another, empirical one dealing with the statistical distribution of random measurement results. I am going to show that these two notions coincide, but it will be convenient for purposes of discussion to designate, temporarily, the ones we have been using up to now as "logical probabilities."

One must consider the real experimental conditions under which statistical probabilities are empirically obtained. A long series of individual measurements is performed on identically prepared identical quantum systems Q. A unique measuring device is usually used in all the measurements, which are made one after another. For the purpose of a theoretical analysis, one may also consider parallel measurements with many copies of the measured system Q and of the measuring apparatus M.

The discussion is easier in the second case. A large number N of individual measurements are performed, but they can as well be considered as a bigger experiment with a large system $NQ+NM$. They are independent, a condition one can mathematically express by assuming the initial state of $NQ+NM$ to be the N-th tensor product of the initial state of $Q+M$ we used previously.

There is little change in the description of the situation by means of histories, except, of course, that the results and the data are N times more numerous. One can then show that the results behave as independent random events from the standpoint of logic. What does it mean? In the usual language of physics, random independent events are characterized by the following properties:

1. Every one of them can occur in a specific trial (i.e., the measurement by apparatus number k).
2. There is no correlation between two different trials: if event n occurs in trial k, this has no influence on the occurrence of another or the same event n' in another trial k' (e.g., when the measurements are made successively, a given result does not affect the next one).
3. The statistical probability P_n of a specific result (finding the eigenvalue a_n for instance) is the limit of the ratio N_n/N when N tends to infinity, N_n being the number of times when the result n occurs. This rather abstract definition (involving the unpractical limit $N \to \infty$) can be made more practical by using statistical significance tests, which I will not discuss.

One can then show that the theory predicts this random behavior of events when the language describing them is our "universal" one with its own logic. Property 1 is obvious, or at least our language envisions it naturally as different histories that must be introduced in a complete family. Since the whole set of N trials can be considered as a unique wider measurement, the consistency of the relevant family of histories follows.

Because of the symmetry properties of a tensor product, one finds that the logical probability for result 1 to be obtained N_1 times, result 2, N_2 times and so on, is given by $p_1^{N_1} p_2^{N_2} \ldots$, each p_j being given by Born's formula. It does not depend on the order of occurrence of the results. This implies immediately the property 2 of randomness when taking the consistency of the relevant family into account: Let a and b be two histories where the same series of events are seen, namely N_1 times the result 1, N_2 times the result 2, and so on, although with a different distribution among the N devices M. One may also consider a history $a + b$ for the two histories occurring one after another in a series of $2N$ trials. Consistency together with the formula for each probability then gives $p(a+b) = p(a)p(b) = p(a)^2$. This is property 2 in a more abstract language.

One can then logically use the notion of a random event for the result of a definite measurement. With it, then, comes the problem of evaluating the corresponding statistical probability. One introduces for each history the frequency of event n as the ratio N_n/N. The coincidence of this frequency (at the limit of large N) with Born's probability p_n can be shown to be a logical consequence of quantum theory as defined on the principles in Chapter 8, even though one may remember that these principles assumed nothing about randomness.

One may introduce for this purpose an arbitrary small number ε and the property $|N_n/N - p_n| < \varepsilon$. A little reflection shows that this is effectively a property in the quantum sense of the word for the large system we are considering. It therefore has a well-defined logical probability $p(|N_n/N - p_n| < \varepsilon)$. A standard application of the central limit theorem of probability calculus (i.e., the so-called law of large numbers) can then be used to show that

$$\text{limit}_{N \to \infty} p(|N_n/N - p_n| < \varepsilon) = 1. \qquad (21.9)$$

This means that the property 3 of random events is predicted with a statistical probability of events coinciding with Born's fundamental formula. In the present approach this is a logical consequence of the theory of interpretation.

One may be somewhat puzzled by this result, which is the last surprise interpretation holds in reserve for us. What does it mean? Could it be, as the method we used seems to suggest, that quantum randomness is a logical necessity deriving directly from the principles stated in Chapter 8? What a feat it would be!

To such an excess of confidence in theory (the besetting sin of theorists), one should prefer, in my opinion, a sounder empirical standpoint. The random character of measurement events is well established empirically, and the order of priority between the two notions of probability should be inverted: empirical randomness and its probabilities give an empirical meaning to logical probabilities. The ideal markers we introduced in Section 132 for signaling a definite history are in agreement with such an approach. In other words, interpretation can be considered in two ways: from abstract principles through convenient histories toward empiricism, or from empiricism to ideally measured histories and logical probabilities.

In any case, histories have done their job with the results in the present chapter. One can willingly be a matter-of-fact empiricist and forget about histories, even though the usual language of physics is their abode. Are they a theorist's device with no direct practical use? It does not matter, since one now understands.

SUMMARY

Measurement theory is a deductive construction relying only on the basic principles of quantum mechanics. Decoherence is its main ingredient. Its two main results are the Born formula for the probability of the measurement outputs and the reduction formula in the case of two successive measurements. The reduction formula (with no specific reduction effect) can be given a general form.

The consistency of interpretation is crowned by the necessary coincidence between the probabilities underlying logic and the empirical probabilities describing randomness in the measurement data. The universal language of interpretation cannot agree with anything other than empirical randomness.

22

Experiments

A book on interpretation can only conclude with some experiments. There is presently renewed interest in analyzing the various aspects of interpretation and submitting it to more and more demanding tests. Unfortunately, it will be impossible to report more than a few of them because their subtlety and their use of elaborate techniques make each of them a topic in itself.

Only three experiments of consequence are described. They were chosen for the evidence they provide on some notions that are particularly challenging to common sense: decoherence, long-range quantum entanglement, and quantum jumps. A few more significant experiments are mentioned in Notes and Comments.

DECOHERENCE

186. The decoherence effect resisted detection for a long time: it is too efficient and observation always occurs after the action, which is too late. One must use a mesoscopic system, somewhere between the Quantum and the Classical.

The present experiment (Brune et al., 1996) is very much like a quantum measurement under conditions like those of Schrödinger's cat. An atom is prepared in a quantum superposition of two states with different quantum numbers: this is the system to be measured. An electromagnetic wave in a cavity affords a collective observable, its phase acting like a pointer on a dial. The average number of photons in the wave controls the mesoscopic character of the wave, between zero and ten or so degrees of freedom. The cavity walls are the environment.

The Superposed State of Two Atom Levels

187. One uses a rubidium atom, which has a unique electron with quantum numbers (n, l, m) outside the complete shells. Two levels $|1\rangle$ and $|2\rangle$ enter into the experiment, the atom being prepared in the

superposed state

$$|3\rangle = \frac{1}{\sqrt{2}}(|1\rangle + |2\rangle). \tag{22.1}$$

Preparation begins by producing the state $|1\rangle$, which is a Rydberg state with high quantum numbers: $n = 50$, $l = m = 49$. It is almost classical and its wave function is confined inside a narrow torus with a large radius (n^2 times, that is, 2,500 times the radius of a hydrogen atom). It is prepared from the rubidium ground state through a series of electromagnetic absorptions with very precise frequencies. This requires a good control of the atom velocity v so that the atom position is known at every instant of time with a good precision.

The state $|2\rangle$ is another Rydberg state with quantum numbers $n = 51$, $l = m = 50$, the difference in energy with $|1\rangle$ corresponding to a frequency $v_{12} = 51{,}099$ GHz. One may notice that an electromagnetic transition between $|1\rangle$ and $|2\rangle$ is made easy by the large value of the transition dipole moment D_{12}, which is proportional to n^2.

In order to prepare the superposed state 3, the atom must cross a (circularly polarized) electromagnetic field oscillating at the resonance frequency ω_{12}. The field can generate only transitions between the two states 1 and 2 and it is contained in a cavity R_1 with length L_1. This length is chosen so that a transition $|1\rangle \to |3\rangle$ is obtained when the atom has crossed the cavity.

This method was invented by Ramsey and the theory is given in Cohen-Tannoudji (1977, Chapter 13, Complement C). The state of the atom is at every moment of time a superposition of the two states 1 and 2. If the electric field is $E = E_o\cos(\omega_{12}t - \beta)$, the coupling atom-field Hamiltonian W has only matrix elements between the states 1 and 2, given by $W_{12} = D_{12}E$. One can easily solve the time-dependent Schrödinger equation for this two-level system under an oscillating force and the atom state is found to be $c_1(t)|1\rangle + c_2(t)|2\rangle$, with $c_1(t) = \cos(|W_{12}|t/\hbar)$ and $c_2(t) = \exp(-i\beta)\sin(|W_{12}|t/\hbar)$, where we do not write the value of the phase β. Choosing L_1 so that $|W_{12}|L_1/v\hbar = \pi/4$, an atom entering the cavity in state 1 leaves in the state

$$\frac{1}{\sqrt{2}}(|1\rangle + \exp(-i\beta)|2\rangle). \tag{22.2}$$

An Ideal Pointer

188. The atom enters then a second cavity C. The analogy between the electromagnetic field in this cavity and a pointer is based on the notion of *coherent states* for a cavity mode, which I will now introduce.

In classical electrodynamics, a cavity mode with frequency ω is strictly equivalent from the standpoint of dynamics to a harmonic oscillator.

The conjugate variables (q, p) for the oscillator are proportional to the magnetic and electric fields, respectively. This is again true in quantum electrodynamics with two nondimensional conjugate observables (q, p) such that $[q, p] = iI$ after a convenient choice of units. The annihilation and creation operators are given $a = (1/\sqrt{2})(q + ip)$ and $a^\dagger = (1/\sqrt{2}(q - ip)$.

Let us first consider a macroscopic field with a large number of photons n (which is not to be confused with the principal quantum number of the atom and will not be mentioned again). The number n is the average value of the operator $a^\dagger a$. All of the average quantities describing the field are large, particularly the average values of q and p, which are to be denoted by (q_o, p_o). The annihilation operator also corresponds to a classical complex and large quantity $a_o = (1/\sqrt{2})(q_o + ip_o)$. Its absolute value is $\sqrt{n}$ and its phase ϕ is that of the field $(E \sim \cos(\omega t - \phi))$. It should be stressed that this is the usual phase, classically well defined.

One can prepare a coherent quantum state of the field in the following manner: A macroscopic field is produced in a cavity S (S for "source") as before. The cavity C is connected to S by an attenuating waveguide (with a cutoff frequency below ω; the field decreases exponentially along such a guide). One would say classically that the complex number a_o is reduced by attenuation with no change of phase (or a well-defined change when the guide impedance is taken into account). When the attenuated field is very weak, quantum effects must be introduced (Figure 22.1).

The modifications are trivial. It can be shown that the oscillator representing the field in cavity C is an eigenstate (coherent state) $|\alpha\rangle$ of the (non-self-adjoint) annihilation operator a, with a (complex) eigenvalue α having the same phase as the source field. One has more exactly $a|\alpha\rangle = \alpha|\alpha\rangle$, the number α being given by $\alpha = Ka_o$, where K is the classical attenuation factor. The average number of photons in cavity C is $n = \sqrt{|\alpha|^2}$. It is sometimes convenient to represent α in the complex plane, the uncertainties on q and p being represented by a circle with unit radius centered at point α (Figure 22.2).

One may think of a circular dial with a radius $\sqrt{n}$ on which a pointer indicates the phase of the generating wave, which may be taken as $\phi = 0$, that is, an initial neutral position of a measuring device.

The Ideal Measurement

189. I will now show that when the atom crosses the cavity C, this is equivalent to a quantum measurement testing whether the atom is in state 1 or 2. This is an ideal von Neumann measurement resulting in a

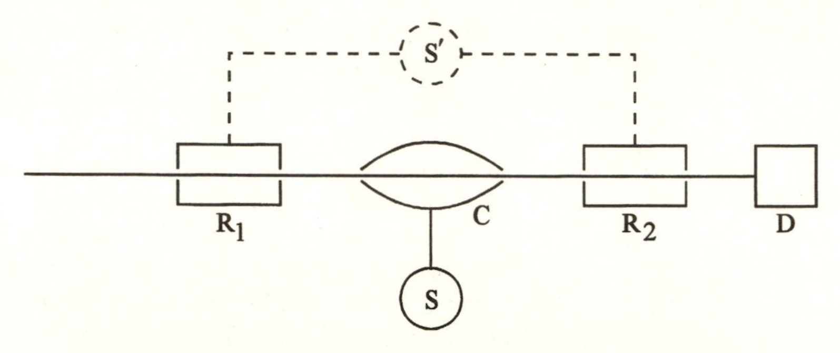

Figure 22.1. Detection of decoherence. An atom in a state of superposition crosses a series of three cavities. The two cavities R_1 and R_2 contain intense oscillating fields with the same phase (the field are produced by the same source S'). The cavity C contains a coherent state of the field with a small average number n of photons. The phase of this field plays the role of a pointer. Decoherence is due to the interaction of this field with the walls of C, which are superconductive mirrors, and also to losses through the holes. The detector D tests the state of the atom.

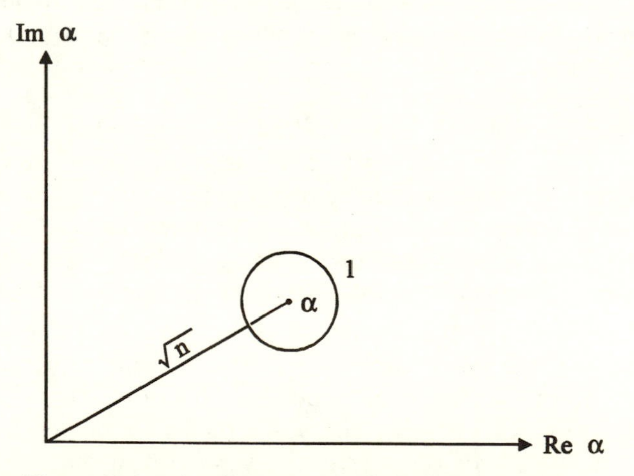

Figure 22.2. A coherent state of an electromagnetic mode is represented by a complex number α whose length is the square root of the average number of photons. The circle with radius 1 represents the effect of the uncertainty relations.

superposed state of the system "atom + field" as in the case of a Schrödinger cat.

Let us first consider a hypothetical case where the atom enters the cavity in the state 1. The field in the cavity has a frequency not far from resonance but sufficiently different for producing an effect quite distinct from what we found in the first cavity R_1. As a matter of fact, the atom remains in state 1 when leaving the cavity, while the state of the field, still a coherent state, has suffered a definite change of phase Φ.

Let the field frequency be $\omega = \omega_{12} + \Delta\omega$, $\Delta\omega$ being larger than the so-called Rabi frequency $|D_{12}|\sqrt{n\,\omega_{12}/\hbar\,\varepsilon_o V}$, where V is the cavity volume and ε_o the vacuum permeability in MKSA units. A straightforward calculation using quantum electrodynamics shows that the atom + field state after interaction is $|1\rangle|\alpha e^{i\Phi}\rangle$ with

$$\Phi = \frac{1}{\hbar}\cdot\frac{L}{v}\cdot\frac{\omega}{\Delta\omega}\cdot\frac{|D_{12}|^2}{2\varepsilon_o V}, \tag{22.3}$$

where L is the length of the cavity. One should particularly notice that this phase change is entirely under control. (For a derivation of these results, see Haroche and Raimond (1994), particularly Section IVB).

If the atom were to enter in the state 2, it would also leave the cavity in state 2, the field phase changing by an angle $-\Phi$. Since the atom enters in the state 3, the outgoing state of the system atom + field is, leaving aside the phase β in equation (22.2),

$$\frac{1}{\sqrt{2}}(|1\rangle|\alpha e^{i\Phi}\rangle + |2\rangle|\alpha e^{-i\Phi}\rangle). \tag{22.4}$$

Decoherence

190. The field in cavity C interacts with an environment that consists of the cavity walls, and it can therefore be affected by decoherence. The aim of the experiment is to see this effect when it is acting. It must therefore be a slow decoherence, which implies a very small dissipation, hence the choice of superconducting walls composed of niobium and kept at low temperature. The cavity is made of two concave mirrors facing each other with a space between them to let the atoms pass through. Damping is determined by the *quality factor* Q of the cavity, corresponding to a damping coefficient $\gamma = \omega/Q$. It turns out that the decoherence coefficient is in the present case proportional to the average number n of photons, as a consequence of the mesoscopic

character of the field. This is why decoherence can be controlled and made weak or strong at will.

The decoherence effect cannot be computed with the help of equation (17.3) because there is no microstable basis in the present case. One must use the general equation (18.10), or as was done in practice, use an oscillator model for the environment. This model was justified for a superconducting environment by A. O. Caldeira and A. J. Leggett (1983b, Appendix C), and losses through the cavity holes can also be treated with similar results. One may work with an (overcomplete) basis consisting of coherent states of the field, and the nondiagonal terms in the reduced field state operator are found to decrease in time with a factor $\exp(-2n \sin^2\Phi.\gamma t)$, replacing in the present case the exponential factor in equation (17.8).

If decoherence were complete, one would have simply an ordinary measurement with no interferences. With the limited amount of decoherence in the present case, one may vary the phase Φ and the average photon number n to preserve or destroy interferences.

Realizing Interferences

191. We cannot observe the quantized field in the cavity C and must therefore resort to the atoms for observing interferences and eventually seeing them disappear. What kind of interferences? They are the so-called *Ramsey interferences*, which are more simply explained when the cavity C is absent (or contains no photon). These interferences are obtained when the atom crosses identical resonating fields in two cavities R_1 and R_2, which are both fed by a common source S' so that their fields have the same frequency and the same phase. These are classical fields, and correcting what I said previously, their frequency is not strictly equal to the resonance frequency but close to it, say $\omega' = \omega_{12} + \Delta\omega'$.

One obtains Ramsey interferences by letting the frequency ω' vary, and if T is the time spent by the atom for going from one cavity to the other, one finds that the probability for finding the atom in state 1 after crossing both cavities is an oscillating function $\cos(\Delta\omega' T/2)$. Notice that $T = L_{12}/v$ is known, L_{12} being the distance between R_1 and R_2. To detect whether the atom is in state 1 or 2 at the end of the experiment, one can use the energy difference between the two states by subjecting the atom to an electric field just above (or below) ionization for state 1 (or 2). State 1 gives an ionized atom that can be detected whereas state 2 goes undetected. Ramsey interferences can then be seen. One may imagine the effect as if the atom were going from R_1 to R_2 along two superposed paths 1 and 2... in Hilbert space.

The Effect of Decoherence

192. Keeping in mind this image of the atom as following two paths together (just like some Schrödinger cat in the path of death and the path of life), one can now see what the effect of cavity C will be. It is a von Neumann ideal measuring device when there is no decoherence, and hence, as we saw in Chapter 19, it does not spoil interferences. On the other hand, when there is full decoherence, there is a real measurement and interferences are destroyed. Since one can make decoherence as weak or strong as one wishes by varying the parameters of the experiment, the effect can be seen (Figure 22.3).

A more precise analysis predicts that the decoherence damping of interferences involves a time Δt during which decoherence acts, which is the time necessary for the atom to leave C and become detected. One may notice that decoherence acts primarily on the field in C, but since the states of the field and of the atom are entangled, the effect is just the same on the atom.

A better test consists in sending an atom first, then a second one a short time Δt later. The second atom (in state 3) finds the cavity field in a pure state superposing two phases Φ and $-\Phi$ if there is no decoherence, or in a mixed state if there is. It therefore tests the decohering

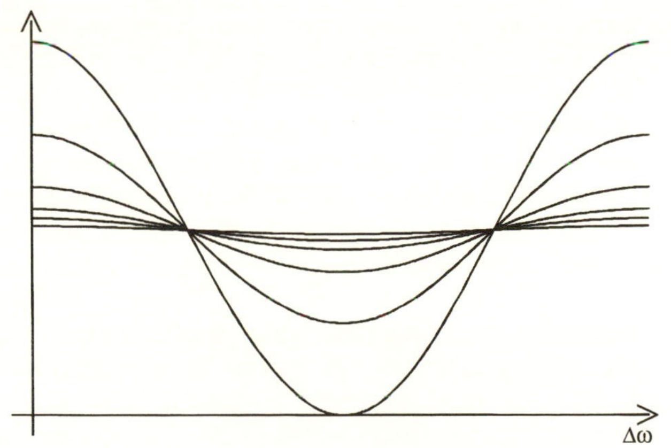

Figure 22.3. The expected Ramsey fringes with an average number of photons ranging from 0 to 5, in a special case of the parameters. The experimental data correspond to very small error bounds.

state of the field. Its component in state 1 interacting with the field component with phase Φ (respectively, $-\Phi$) leaves the field in a superposition of the two phases 0 and 2Φ (respectively 0 and -2Φ). There are two different ways of getting phase 0, and therefore one sees correlations between the two atoms if these two processes in a pure state, but no correlation if they are mixed incoherently.

I will not enter into a more detailed discussion of the theory and the results, except for insisting on the fact that every parameter can be controlled and checked through independent measurements. The agreement between theory and experiment is excellent. One can therefore conclude that the decoherence effect is observed in this experiment, both when it is weak, as in an elementary quantum system, and when it is strong, as usually happens in a macroscopic framework—and also when it is at intermediate strengths.

DISTANT ENTANGLED STATES

193. An entangled state is a quantum superposition of two distinct physical systems. This is a very frequent situation because any composite system whose wave function is not simply a product of the wave functions of its components is entangled. The existence of these states is proclaimed by the Pauli principle, and in that sense, it is responsible for a host of physical properties from the hardness of a stone to the laser. In spite of that, our imagination and common sense are tempted to reject some of their consequences, particularly when they imply very long-range correlations. The experiment I am going to discuss (Aspect, 1982a, b) confirms that entanglement exists, so the best we can do is to accept it.

Einstein, Podolsky, and Rosen (1935) were the first to call attention to some peculiarities of entangled states in their famous paper on the notion of reality in the atomic world. Without returning to a topic we have already considered, a few assumptions that are often compared and opposed when discussing the problem of reality may be mentioned:

1. The predictions of quantum mechanics are correct.
2. No signal, no interaction can propagate faster than light.
3. When two objects (e.g., two particles) are far from each other, one can consider the properties of one of them without having to mention the other.

The last assumption, which looks so much like a matter of common sense, is called *separability*. Bohr's answer to the EPR elements of reality can be summarized in a short sentence: quantum mechanics is not separable.

Bell revived the question in 1964, bringing it from the realm of ideas to the domain of experiments. He made the notion of separability more precise by assuming the existence of local "supplementary parameters" and comparing the consequences of that assumption with conventional quantum mechanics.

One may consider as an example two photons 1 and 2, which are prepared together before separating and going in opposite directions. A dichotomic (yes-no) measurement is performed on each of them, one measurement deciding whether photon 1 is linearly polarized in a direction a and the other deciding whether photon 2 is polarized in a direction b. Each measurement consists in letting a photon cross a polarizer (a birefringent crystal) in front of a detector (a photomultiplier) as shown in Figure 22.4.

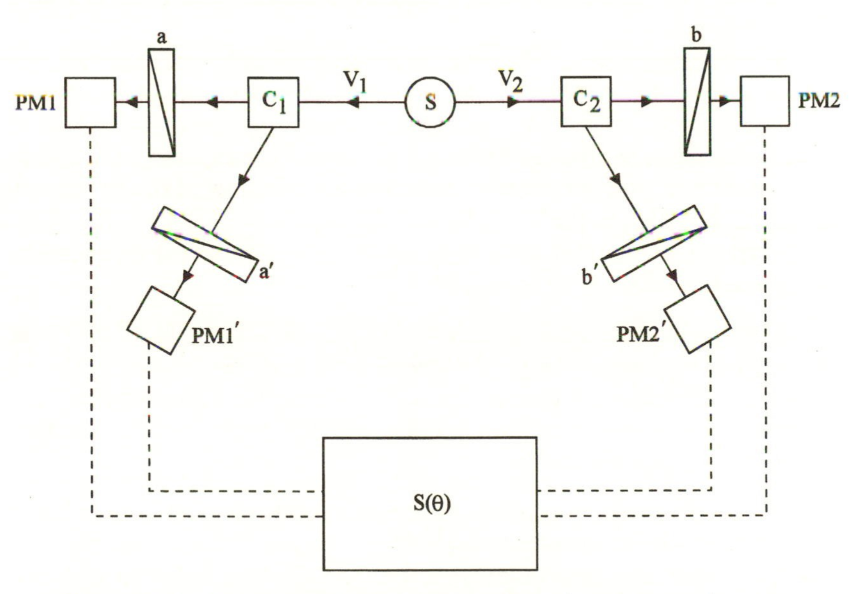

Figure 22.4. Detection of entanglement. Two photons ν_1 and ν_2 are produced in the region S from a cascade decay of an atom. Optical commutators C_1 and C_2 direct the photons toward four photomultipliers $PM1$, $PM2$, $PM3$, and $PM4$. There is a linear polarizer in front of each photomultiplier, the one in front of $PM1$ for instance letting through only a photon with a linear polarization in direction a. The data from the four detectors are sent to a computer, and their statistics are used for computing the Bell function S.

Let one consider two orthogonal space directions x and y in the plane perpendicular to the photon trajectories. One denotes by $|x_1\rangle$ the state of photon 1 polarized in direction x, with similar notations for direction y and photon 2. The two photons will be assumed initially in the entangled state

$$\frac{1}{\sqrt{2}}(|x_1, x_2\rangle + |y_1, y_2\rangle), \tag{22.6}$$

an expression that is valid whatever the two orthogonal directions x and y one may choose (one may notice an unimportant difference in sign with equation (15.1) for the case of two spin-1/2 particles).

The predictions of quantum mechanics for this experiment are as follows:

- Every individual measurement is a random event: "yes," photon 1 is polarized in direction a with a frequency 50%, and "no" in 50% of the cases, with the same predictions for photon 2 and direction b.
- Joint measurements, however, are strongly correlated: If the two polarizers are oriented along the same direction, the only possible pairs of results are "yes-yes" and "no-no" (perfect coincidence). If the two directions a and b make an angle α, one must introduce a correlation function $C(a, b)$ involving the statistics of observation in the ratio $[N(\text{yes-yes}) - N(\text{no-no})]/[N(\text{yes-yes}) + N(\text{no-no})]$, N denoting a number of observed results. The prediction of quantum mechanics is $C(a, b) = \cos 2\alpha$.

According to EPR, the complete correlation for $\alpha = 0$ would mean that there exists really pairs of photons that are polarized in the same direction, whereas quantum mechanics makes no such assumption since it does not even presume real properties. Bell preferred to introduce the following hypotheses:

- A real property belongs in common to the two photons and it is expressed by some parameter λ.
- The result of a measurement of photon 1 with a polarizer oriented in a direction a is expressed by a result function $A(\lambda, a)$, which can take only the value 1 (for "yes") and -1 (for "no"). This means that the result of the measurement is completely determined by the polarizer direction and the real property of the photon system. Similarly, the result of the measurement on photon 2 with a polarizer along direction b is a function $B(\lambda, b)$. This is an assumption of separability because $A(\lambda, a)$ depends on the real property λ and on the polarizer that is present here and now when the

detection of photon 1 is made, and not on some distant condition such as the direction b of the other polarizer.

- The parameter(s) λ vary from one pair to another, which is why individual measurements are random, the values of λ having a probability distribution $\rho(\lambda)$, which is positive and normalized.

Bell's achievement was to draw experimental consequences from these assumptions. One must perform a large number of measurements for obtaining a correlation function, and these correlations must be found for two different orientations (a and a', b and b') for each polarizer. One can then define the following combination of correlations

$$S = C(a,b) - C(a,b') + C(a',b) + C(a',b'). \tag{22.7}$$

Bell's theorem states that his assumptions imply the inequalities

$$-2 \leq S \leq 2. \tag{22.8}$$

But we saw that quantum mechanics predicts $C(a,b) = \cos 2\alpha$. One may then choose the four directions (a, b, a', b') as making a fan whose branches have the same angle θ as in Figure 22.6. Choosing $\theta = \pi/8$, quantum mechanics implies $S = 2\sqrt{2}$, in violation of Bell's inequalities. This is a clear indication of nonseparability.

194. The main experimental difficulty is to prepare a pair of photons in the entangled state (22.5).

A calcium atom in its ground state g is used. It is brought to an excited state e through the simultaneous absorption of two photons with different frequencies ν' and ν'' from the light of two lasers. This two-photon transition is a nonlinear effect demanding extremely well-collimated laser beams falling simultaneously on the atom. Their frequencies must be such that the sum $\nu' + \nu''$ is equal to the transition frequency for the two states

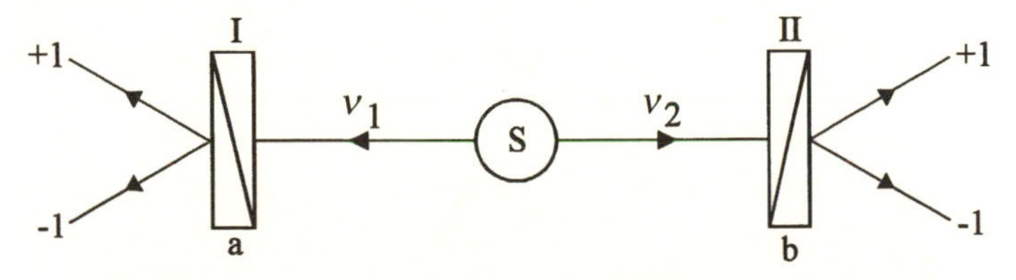

Figure 22.5. The EPR situation. Two photons ν_1 and ν_2 are produced by a source S in a definite entangled state. The two polarizers I and II are respectively oriented along directions a and b. A signal $+1$ or -1 is obtained when the photon crosses the polarizer or not.

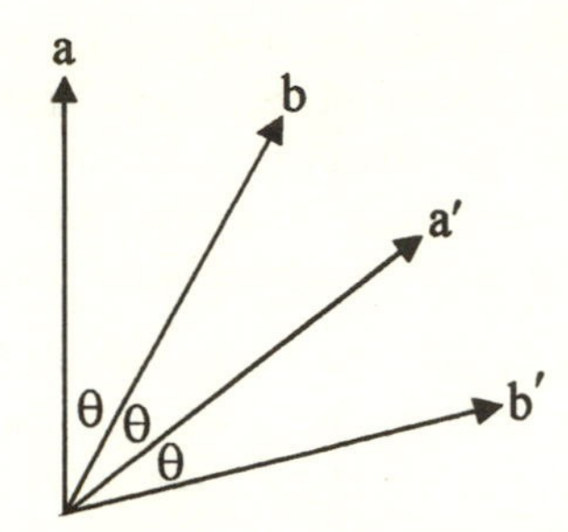

Figure 22.6. The four directions a, b, a', b'.

g and e, with a relative precision which must be at least 10^{-8}. Each atom must also be isolated, a condition that can be satisfied by using an atomic beam in vacuum, the two intense laser beams being strongly focused on the atomic beam. These demanding conditions are the real difficulties of the experiment.

The atom in state e has a cascade decay: It first produces a photon 1 with a frequency ν_1 when falling on an intermediate excited state r, which decays very rapidly while producing a second photon 2 with a frequency ν_2.

The results of the experiment are shown in Figure 22.7, where a violation of Bell's inequalities is clearly seen because S oversteps the bounds -2 and $+2$ for some values of the angle θ. On the other hand, these results are in good agreement with quantum mechanics.

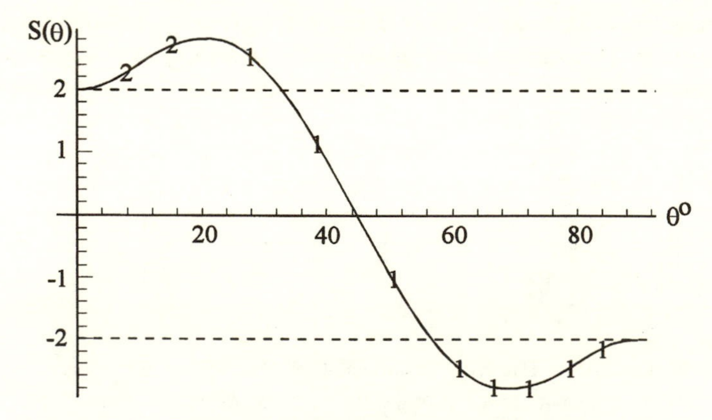

Figure 22.7. The results of the experiment. The function $S(\theta)$ exceeds the bounds ± 2 arising from Bell's inequality, and it agrees with the prediction from quantum mechanics.

After this experiment, several speculative ideas were proposed for saving separability: maybe some information is going from one detector to the other; taking into account the fact that many photon pairs are not detected because of the imperfect efficiency of photomultipliers, it might be that the mechanisms responsible for this inefficiency conspire with the hidden parameters to reproduce quantum predictions; etc. These wild constructions are so far from the simplicity and beauty of Bell's assumptions that they are much more unbelievable than quantum mechanics. They would not recover intuition if they could cope with realism, and it is much simpler to accept the evidence: intermingled states of distant particles exist and this existence belongs to reality.

OBSERVING A UNIQUE ATOM: HISTORIES AND QUANTUM JUMPS

Trapped Atoms

195. The possibility of continuously observing a unique atom could not even be envisioned in the early years of quantum mechanics, and standard books often denied it explicitly. The interest in this kind of experiment is threefold: it shows a quantum history as it unfolds in time; it sees quantum jumps; and it transforms the random character of quantum mechanics into a random signal rather than a statistical collection of measurements on different atoms.

The experiment is possible because one can maintain a unique atom inside a "trap." This feat has been realized with neutral atoms as well as ions. An ion can be enclosed for instance in a "Paul trap," which is shown in Figure 22.8. It is a metallic box whose top and bottom are the sheets of a two-sheeted hyperboloid, the side of the box being a one-sheeted hyperboloid. Top and bottom are submitted to a time-varying electric potential $V = U_o + U_1 \cos(\omega t)$, while the side is at potential $-V$. Through a convenient choice of the static (U_o) and oscillating (U_1) potentials as well as the frequency ω, one can exert an electric force on an ion, which keeps it trapped in the box for a long time, the record being ten months!

Many experiments on interpretation have been performed with trapped ions or neutral atoms, but I will discuss only one of the first and simplest ones.

Observing a Unique Atom

196. A unique barium ion is trapped. It is illuminated by a blue laser beam whose frequency resonates with the transition from the ground

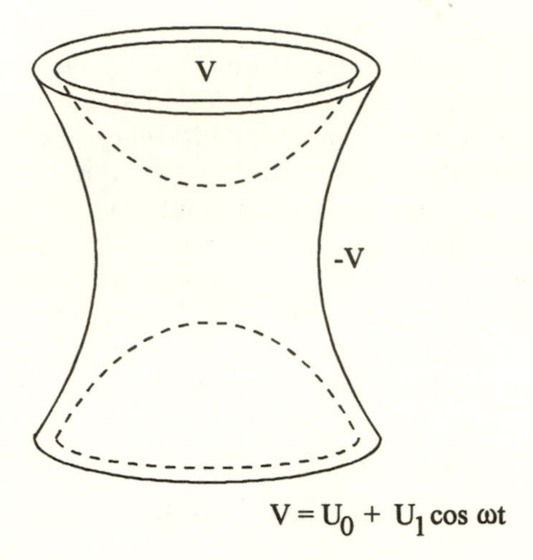

Figure 22.8. An ion trap. It consists of a box whose top, bottom, and sides are sheets of hyperboloids maintained at oscillating electric potentials. This device can confine an ion for a long time. Other devices using lasers now exist for neutral atoms.

state 1 to an excited state 2. If subjected to a laser beam of low intensity, the atom in state 1 can reach state 2 by absorbing a photon. When in state 2, the atom can decay and go back to state 1 by emitting a photon. An atom happening to be in state 2 when no laser is acting would decay *spontaneously*, emitting a photon in an arbitrary direction: this phenomenon is fluorescence. When a laser beam is present, the probability for an emission of the photon in the direction of the beam is multiplied by a factor $(n + 1)$, n being the number of photons in the laser mode: this phenomenon is stimulated emission.

When the laser beam is intense and therefore n very large, most emitted photons are stimulated and they get lost in the laser beam. A small fraction, however, is emitted spontaneously in other directions and these photons can be detected. The laser action, moreover, implies an acceleration of the process, since the cascades

$$1 \rightarrow 2 \rightarrow 1 + \text{photon in a laser mode or a fluorescent mode} \quad (22.9)$$

become more rapid when the laser intensity increases. As a consequence and although the fraction of spontaneous emission is small, the rate of production of these photons can be high, some 1,600 fluorescence photons per second in a typical run, enough for taking a photograph or seeing the atom with the naked eye as a blue luminous point!

One may notice that the cascade (22.9) as I wrote it is not a consistent history and therefore not a rational description of the phenomena. This can be seen in a model where the electromagnetic field has a unique mode, which is the laser one. Solving the Schrödinger equation for the system atom + field, one finds the total state to be a linear combination of a state $|1, n\rangle$ with the atom in state 1 and n photons in the field, and a state $|2, n-1\rangle$. This is a "dressed state" of the atom, which is dressed in a photon garment. When inserted into histories representing the cascade (22.9), the cascade description of dynamics contradicts consistency.

The Fluorescence Signal

197. The fluorescence signal consists of photons that are emitted outside the laser mode, and it can be observed with a battery of photodetectors. Our aim will be to interpret it, that is, relate it to the quantum formalism and find the information it contains. Consistent histories provide the best framework for this analysis.

Which histories? The simplest approach for finding them is probably the following one: Let us assume for simplicity that the atom stands motionless at a point O. One introduces a spherical shell S centered at O with a radius R and a width ΔR, the two caps through which the laser beam is going are cut out as in Figure 22.9. We shall then build up histories where every property can be expressed by "a fluorescence photon is in S at time t_i."

The photon localization property makes sense, as already mentioned (see Notes and Comments). It can be expressed, as usual, by a projector

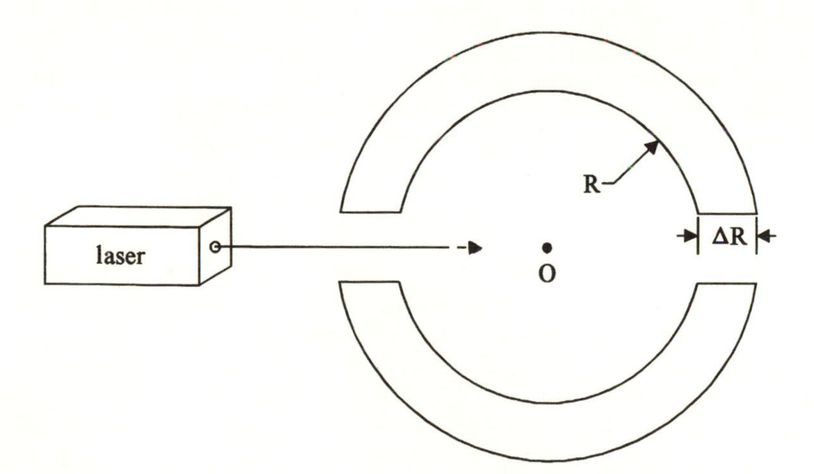

Figure 22.9. The spherical region S described in the text.

in Hilbert space. One may notice that this property properly specifies a fluorescence event because a photon spends a time $\Delta R/c$ in the shell S, thus defining the latitude one has in the choice of t_i for every event. When such a property occurs, it will be immediately followed by a photon detection transforming it into a datum. The history language is therefore very close to the language of observation and more sensible than if one were saying "the photon is emitted at time $t_i - R/c$," for various reasons I will not enter into.

The fluorescence emission is therefore described by histories, each of them stating the times $(t_1, t_2, \ldots t_n, \ldots)$ when a photon crosses the shell S, with $t_1 < t_2 < \cdots < t_n$. One may notice that it looks like a family with continuous values for the times, which I have avoided discussing until now. Rather than making the theory more general (and more complicated), one may simply remain in a known framework by means of a simple trick: I will consider discrete values of time, like the tick-tock of a clock, and every time t_i is written as $k_i \Delta R/c$ where the numbers k_i are integers. One thus obtains a family of histories where each history is prescribed by a series of integers $(k_1, k_2, \ldots k_n, \ldots)$.

These histories do not mention anything about the exact number of photons in the laser beam, and contrary to the cascades (22.9), they are consistent. This is because two distinct histories must differ at least at one time t_i or an index k_i for which one history "sees" a photon in S and the other does not. The corresponding projectors are orthogonal and the GMH consistency conditions are satisfied. As a matter of fact, a careful study of the consistency conditions would show corrections of order $\hbar/(E_2 - E_1)\Delta t$, Δt being the average time elapsing between two fluorescence emissions in the same quantum mode. This is completely negligible, however, and we are back to our familiar framework.

Note. Photodetectors are usually far from being one hundred percent efficient, and one could therefore distinguish between emitted photons and detected photons. This point will be considered as inessential.

Probabilities

198. A significant feature of this experiment is to provide a time signal rather than a collection of independent measurements. A fluorescence signal is given by a series of times or time indices $(k_1, k_2, \ldots k_n, \ldots)$, which are separated on the average by a large interval $\Delta k = cT/\Delta R$, where T denotes the average time interval between two arrivals of a photon in the shell S. Every history has a well-defined probability in view of consistency, and one can then consider it as a random signal, which is a well-known notion in signal theory. The calculations are

somewhat involved because one must take into account the atom-laser coupling in the evolution operator, but the result is simple: one obtains a Poisson signal for fluorescence, the time interval t between two photon observations being a random variable with a distribution probability $\exp(-t/T)\,dt/T$, up to negligible corrections for very small values of t.

Quantum Jumps

199. When Heisenberg invented matrix mechanics, he relied on the notion of quantum jumps as they had been introduced by Bohr: an atom can jump suddenly from a state 2 to a state 1 with a photon emission and one cannot observe a continuous transition between the two states. This idea is fully confirmed by an analysis using histories for logical consistency.

No laser being present, it is convenient to introduce a time t_1 when the atom is in state 2 (or more exactly the property $|2\rangle\langle 2|$ holds) and a time t_2 when the property $|1\rangle\langle 1|$ holds. In order to see what quantum jumps mean, one can also introduce an intermediate time t' when some transient property holds, which must of course be made precise by consistency.

The results of this analysis are as follows (Omnès, 1994a, Sections 5.8 and 11.1): Consistency is possible only if $t_2 - t_1 \gg \Delta t \equiv \hbar/(E_2 - E_1)$, E_1 and E_2 being the atom energies in the two states 1 and 2. The intermediate properties have no transient character. They can only state that the atom has still property 1 or already property 2, under the subsidiary conditions $t' - t_1 \gg \Delta t$ in the second case and $t_2 - t' \gg \Delta t$ in the first one. Quantum jumps are therefore logically unavoidable.

Observing Quantum Jumps

200. One can directly observe quantum jumps in the present experiment. A barium ion has an intermediate level 3 between the two previous ones, with $E_1 < E_3 < E_2$. The quantum numbers of these levels are such that the two transitions $2 \to 3$ and $3 \to 1$ are forbidden, which means in practice that the corresponding matrix elements in the Hamiltonian are very small, so these events must be very rare. As a matter of fact, a transition $2 \to 3$ is so rare that it is practically never seen in normal conditions and even more so when the blue laser imposes forced transitions to state 1.

The situation is changed when a red laser at the resonance frequency for transitions $2 \to 3$ is acting. From time to time, there will be a stimulated decay in which the atom jumps from state 2 to state 3. It

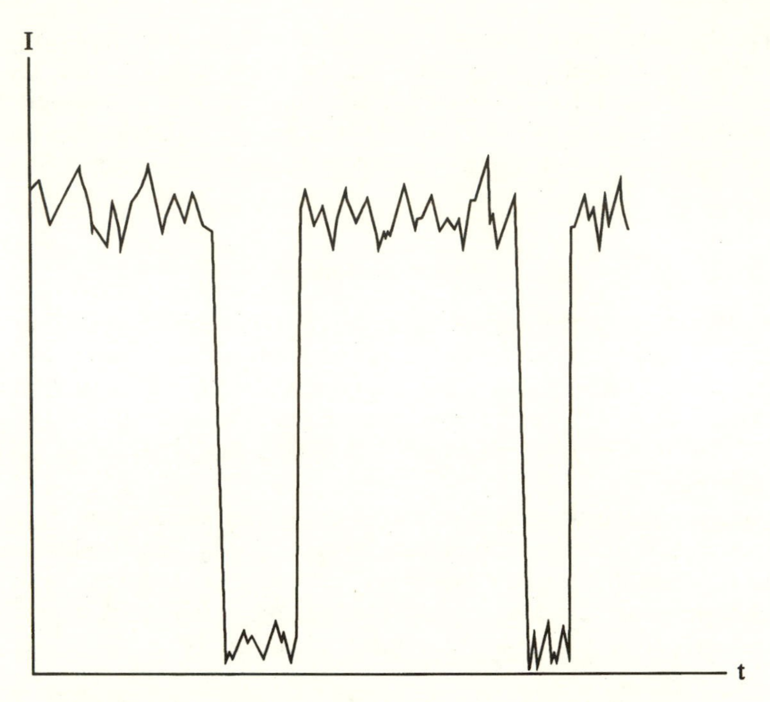

Figure 22.10. A typical observation of quantum jumps using an ion trap.

remains a long time in the intermediate level because the transition $3 \rightarrow 1$ is forbidden. As long as the atom is in state 3, it does not emit fluorescence photons and the fluorescence signal stops. This situation lasts until the decay of state 3 into state 1. The blue game can then start again.

What I have described plainly in words is again a history, or at least a perfectly consistent fragment of a history. It is shown experimentally by a sudden interruption of the fluorescence signal as shown in Figure 22.10, which is followed by an absence of signal, then by a return of fluorescence. Theory gives an excellent account of the statistical properties of the time distribution of disappearances and reappearances of the signal. It should be noticed, of course, that the transient continuous transition one can see in the signal is due to damping in the electronic appliance registering the signal and it has no quantum significance. Finally, one sees only quantum jumps, as expected.

Notes and Comments

PART ONE

Chapters 1 to 4

When writing these historical chapters, I relied mainly on Darrigol (1992), Jammer (1966), and Mehra (1973).

PART TWO

Chapters 5 to 7

This tentative history of interpretation is based on many books and papers. The most useful one is the source book by Wheeler and Zurek (1983). See also Chevalley (1997), Murdoch (1987), d'Espagnat (1976), Jammer (1966, 1974).

Chapter 5

The best documented works on the early philosophical aspects of the Copenhagen interpretation are the book by Murdoch (1987) and the thesis by Catherine Chevalley (1997), which is unfortunately not yet published as a book. See also the classic book by Jammer (1974).

The discussion of the interest shown by Bohr, Heisenberg, and Pauli about a language specific to interpretation is inspired by Chevalley (1997), and it tries to offer a shortened version of similar considerations together with other aspects of Bohr's philosophy by Murdoch (1985).

35. This example was communicated to me by Serge Caser.
39. For the time-energy uncertainty relations, see Aharonov and Bohm (1961), W. Thirring (1979), and P. Pfeifer and J. Fröhlich (1996).
40. The recent rediscovery of Pauli's important philosophical investigations is due to Catherine Chevalley (1997). See Pauli (1936, 1950, 1979). One may notice that Pauli's critique could have been extended to the category of inherence (one of those introduced by Kant), which seems to proscribe the existence of two identical objects and therefore the Pauli principle. I may be wrong, however, on that point since Kant criticized Leibniz' contention against nondistinguishable objects.
41. The complementarity principle is given in Bohr (1928a, b).
43. See Heisenberg (1930, 1958).

Chapter 6

For the question of realism, see Bell (1987), d'Espagnat (1985, 1994), Jammer (1974), Omnès (1994b).

49. Von Neumann's model appears in the last pages of the book by von Neumann (1932), reproduced in Wheeler and Zurek (1983).
53. For the theory of pilot waves, see also Bohm (1966, 1993).
55. See Bell (1964, 1966, 1987), Clauser et al. (1969), d'Espagnat (1976).

Chapter 7

63. For the quantum behavior of some macroscopic systems, see Leggett (1980, 1984, 1987) and Clarke et al. (1988).
67. See also Omnès (1994b).

PART THREE

Chapter 8

One can find various expressions of the quantum principles in textbooks. See Cohen-Tannoudji, Diu, and Laloë (1977); Feynman, Leighton, and Sands (1965a); Laudau and Lifschitz (1967); Merzbacher (1970); Messiah (1995). For a mathematical analysis of Hilbert spaces, see Reed and Simon (1972).

The method of C*-algebras was proposed by Haag and Kastler (1964); path integrals are described in Feynman and Hibbs (1965b). I did not take into account superselection rules in my brief introduction. See Wick, Wightman, and Wigner (1952). It should also be mentioned that Principle 1 is valid, irrespective of EPR correlations.

77. The best introduction to logic, as far as I know, is in the source book by van Heijenoort (1967).

Chapter 9

This chapter is based on von Neumann (1932). See also Mittlestaedt (1978) and Omnès (1994a), Chapter 3.

82. The existence of a position operator in the case of a photon, and more generally a relativistic particle, was considered by Pryce (1948), Newton and Wigner (1949), and Foldy and Wouthuysen (1950). One of their most remarkable results is the nonexistence of such an operator for the massless particles with spin greater than $1/2$, including the photon (and one-component neutrinos). This was established definitely by Wightman (1962). The specific needs of interpretation lead me to show that a projection operator (or rather a set of equivalent projection operators as discussed in Chapter 10) can nevertheless give a definite meaning to the property stating that a photon is in a space domain V, in a given reference frame, under the condition that its wavelength is bounded (Omnès, 1997b).

Chapter 10

The relation between classical and quantum physics remained an active field of research since the initial works by Brillouin, Wentzel, and Kramers in the 1930s (see for instance, Messiah, 1995, Chapter 6). For the continuous expansions on Gaussian wave functions (coherent states), see Hepp (1974); Ginibre and Velo (1979); Hagedorn (1980a, b, 1981).

The most powerful methods were developed in the framework of microlocal analysis. This is unfortunately a rather highbrow mathematical theory and some physicists find its access difficult. See the book by Hörmander (1985). A simplified practical introduction is given in Omnès (1994a, Chapter 6, Appendix A).

88. The introduction of a classical probability distribution $W(x, p)$ corresponding to the quantum density operator was made by Wigner (1932). It was extended to a connection between an observable and a function of (x, p) by Weyl (1950). This Weyl calculus was later included in the framework of microlocal analysis by Hörmander (1979); see also Hörmander (1985, Chapter 18.5). For different approaches to microlocal analysis (or pseudo-differential calculus), see Stein (1993) or Grigis and Sjöstrand (1994).

89 et sq. The projectors that are introduced here are used intensively in the books by Hörmander, Stein, and Grigis and Sjöstrand. Some of their properties, most useful for application to physics, are given in Hörmander (1979). Their adaptation to the specific needs of interpretation was made by the author (Omnès, 1989, 1994a, Chapter 6, 1997a, b). See also A. Voros (1978) for an application of Hörmander's methods to other problems in semiclassical physics.

Chapter 11

The references are the same as in the previous chapter. One must add however the original article by Egorov (1969) and its account by Grigis and Sjöstrand (1994). Another method for deriving classical physics using decoherence and coarse graining is given by Gell-Mann and Hartle (1993). See, in a similar perspective, the book by Giulini et al. (1996), in which one can find many interesting and useful developments.

98. For a lucid description of the Stern-Gerlach experiment, see Feynman (1965b).

Chapter 12

The standard language we translate by histories has always been used in the ordinary practice of physics. The notion of history seems to appear for the first time in a paper by Aharonov, P. G. Bergman, and J. L. Lebowitz (1964), where each property is supposed to result from a measurement in the framework of the Copenhagen interpretation. The notion of history that is used in this book was introduced by Griffiths (1984); see also Griffiths (1987) for an elementary treatment. The logical structure originating in histories was found by the author (Omnès, 1988), who extended the method to classical physics. Another formulation stressing the effect of decoherence was proposed by Gell-Mann and Hartle (1991).

The question of a possible relation between histories and real events was raised in a critical paper by d'Espagnat (1989), and it is discussed in the comments to Chapter 20. The outcome of the ensuing discussion was a better assessment of histories as a universal language, a powerful method for organizing interpretation, and not a reestablishment of standard realism.

102. One can give a more general criterion for exclusiveness in a family of histories, even for two histories having no common reference time. Exclusiveness must then be defined in a roundabout way by using the logical meaning of a history as explained in the next chapters: Let $\{E_k^{(j)}(t_k)\}$ and $\{F_q^{(p)}(t_q)\}$ be the sets of projectors for the two histories. Suppose that a time t can be inserted in the first time series after and index k so that $t_k \leq t < t_{k+1}$, and similarly $t_q \leq t < t_{q+1}$ and there are two projectors E and F such that: (i) The two logical implications $E_k^{(j)}(t_k) \Rightarrow E(t)$ and $F_q^{(p)}(t_q) \Rightarrow F(t)$ hold (ii) E and F are exclusive: $EF = FE = 0$. Then the two histories can be said to be exclusive. This is however a mathematical game because it would be foolish not to use the freedom one is afforded in the choice of reference times to obtain obvious exclusiveness at a common time.

104. The geometrical construction of the operations *and*, *or*, and *not* is given in Omnès (1992, 1994a) in more detail. See also Griffiths (1996, 1998) for another approach.

105. The definition of inference by means of probabilities in quantum mechanics was proposed by the author (Omnès, 1988). The proof of the main theorem in this section is given in Omnès (1994a, Chapter 5, Appendix B).

Chapter 13

112. Gleason's theorem was first given by Gleason (1953). A clearer though still technical proof was given by Jost (1976). For some generalizations, see Maeda (1990).

113. For a proof of equation (13.1) in the case $n = 2$ and a discussion of $n > 2$, see Omnès (1994a, Chapter 4, particularly Appendix B).

Chapter 14

The consistency conditions were introduced by Griffiths (1984). Another form was given by Omnès (1990, 1994a, Chapter 4, Appendix A). Various examples are given by Griffiths (1984, 1987, 1996, 1998) and Omnès (1994a, Chapters 5, 6 and 11); see also Dowker and Kent (1996). The strong GMH conditions were proposed by Gell-Mann and Hartle (1991) from a consideration of decoherence. One spoke during some time of "consistency conditions" (*à la* Griffiths) and of "decohering histories" (*à la* Gell-Mann and Hartle). Time decided in favor of the latter for reasons of convenience (see, for instance, Griffiths [1996] A systematic algebraic method for formulating histories was proposed by Isham and Linden (1994). See also Isham, Linden, and Schrekenberg (1994).

120. One can prove that the GMH conditions in equation (14.4) imply additivity by using for instance the explicit Griffiths (1984) conditions and verifying that they are satisfied as a consequence of equation (14.4).

122. The problem of interpreting interference experiments is discussed at length by Feynman (1965a, Chapter 1). The usual method of reducing the theory of interferences to differences in the length of pathways can be justified from the Schrödinger equation by the WKB method (see Messiah, 1995, Chapter 6; or Landau and Lifschitz, 1968, Chapter 7). In the case of photons, there is no direct derivation from quantum electrodynamics, to my knowledge, though this is a relatively easy exercise. A derivation from the Maxwell equations is given in the book by Born and E. Wolf (1959, Chapter 7).

123. The discussion of "approximate logics" is more explicit in Omnès (1994a, Chapter 5, Secton 9, and Chapter 8, Sections 10 and 11).

126. The interpretation rule was proposed by the author (Omnès, 1988).

Chapter 15

The complementarity principle was proposed by Niels Bohr at a meeting in Como and published in Bohr (1928a, b). These papers are reproduced in Wheeler and Zureck (1983); see also the interesting comments by Rosenfeld and Heisenberg in this source book as well as Chapter 7 in Jammer (1966). The point of view on which the present chapter relies was proposed in Omnès (1992). See also Griffiths (1996, 1998) for a slightly different approach.

126. The no-contradiction theorem was announced in Omnès (1992) and its proof can be found in Omnès (1994a, Chapter 5, Appendix C).

127. The spin-1/2 example was given by Griffiths (1984).

128. The example of a straight-line motion was treated for the first time by Mott (1929a), and this paper is reproduced in Wheeler and Zurek (1983, p. 129).

130. The example of a spin 1 is adapted from Aharonov and Vaidman (1991), in a form proposed by Kent (1997).

132. The relation that is proposed in this section between histories and systematic von Neumann measurements is new.

The relation between this Heisenberg-like approach and consistency conditions is as follows: One assumes the imaginary measurements to be *ignorable*. This means that the probability of the last measurement is the same, whether or not some imaginary measuring devices are acting but their results are ignored (i.e., one never looks at them or, mathematically, one traces them out). By "some" devices, one means any number of them, arbitrarily chosen from none to all of them, except, of course, for the last one (which is real or treated as real and is always acting). This condition of ignorability entails the Griffiths consistency conditions if the devices (except the last one) are of the ideal von Neumann type. One obtains the GMH conditions if they are "realistic," that is, interact with an environment fixing their data by decoherence. These results, which are easily established, may give a more physical flavor to the history approach for some people who have a strong inclination for the Copenhagen interpretation.

133. The "EPR" experiment was published by Einstein, Podolsky, and Rosen (1935), (reproduced in Wheeler and Zurek (1983, p. 138). Bohr's answer was immediate (1935a) and later developed in Bohr (1935b), the two references being also reproduced in the book by Wheeler and Zurek. A more convenient version of the EPR experiment using spins was proposed in the book by Bohm (1951, Chapter 22) (reproduced in Wheeler and Zurek, p. 356). There is a vast literature on this topic; for an introduction, see Selleri (1994). The necessary calculation of consistency traces and probabilities that are referred to in the present section is given in Omnès (1994a).

Some authors keep seeing a mysterious instantaneous exchange of information at long distance when the two measurements I mention in the text are made along the same direction ($n = n'$). They seem not to appreciate the fact that two previous elements of information must be available for the realization of these conditions: (*i*) an initial state involving spin correlations; (*ii*) a prior decision or information ensuring that the two spin measurements will be performed along the same direction. When this is taken into account, there is never a transmission of information nor a real prediction but only correlation in a pre-set experimental arrangement.

Chapter 16

The present conception of determinism does not claim novelty. It was proposed in its present form in Omnès (1990, 1992, 1994a).

137. For the compared statistics in classical and quantum physics in the case of a chaotic system, see Omnès (1994a, Chapter 10). This specific topic has been little investigated in spite of a vast literature on quantum chaotic systems.

140. The consequences of a reversal of premises is discussed from a wider philosophical standpoint in Omnès (1994b).

Chapter 17

The idea that macroscopic quantum interferences could be destroyed by the environment is old (see, for instance, Heisenberg, 1930, 1962), but the mechanism was not clear. A noise mechanism was considered for instance by Daneri, Loinger, and Prosperi (1966). The modern idea of decoherence is due to van Kampen (1954) and Zeh (1970).

Theorists had an impression that the detailed control of the phase of a wave function for a large system would be difficult, and they therefore resorted to models. The most investigated one uses a representation of the environment by harmonic oscillators: Hepp and Lieb (1974); Leggett (1987); Caldeira and Leggett (1983a); same authors (1983b, erratum Ann. Phys. (NY), **153**, 445 (1983)); Unruh and Zurek (1989); Hu, Paz, and Zhang (1992); Paz, Habib, and Zurek (1993); Paz (1994). Many of these calculations rely on an explicit summation over Feynman paths that was given by Feynman and Vernon

(1963). A pedagogical introduction to these results is given by Zurek (1991), and Zurek's ideas (1981, 1982) in the research on decoherence are important. See also Emch (1965) and Lindblad (1979).

A stochastic model of "quantum diffusion" was studied by Gisin and Percival (1992, 1993); Halliwell and Zoupas (1995); Scheck, Brun, and Percival (1995). Another important model was proposed by Joos and Zeh (1985): decoherence is due to the accumulation of phase shifts in the collisions of a macroscopic object with the molecules or photons of an external environment.

An insightful discussion of the physics of decoherence is given by Gell-Mann (1994). A review can be found in the book by Giulini et al. (1996). For the experimental evidence on the effect, see the comments to Chapter 22.

Chapter 18

153. For a thorough discussion of the projection method in the theory of irreversible processes, see Balian, Alhassid, and Reinhardt (1986).

154. The theory given in this section and the following ones was proposed by the author (Omnès, 1997c). Its relation with the more general projection method was suggested by Roger Balian.

156. The importance of the diagonalization basis for decoherence was stressed by Zurek (1981, 1982). The arguments we present in favor of a basis in three-dimensional space are close to those of Zurek (1994, p. 175) and of Gell-Mann and Hartle (1993).

For a derivation of the Fokker-Planck equation from the oscillator model, see Caldeira and Leggett (1983a).

Chapter 20

For a different appreciation of the "fundamental" questions in this chapter, see, for instance, Bell (1987), d'Espagnat (1985, 1994), and Wigner (1976). The relevance of Borel's axiom for the questions occurring in this chapter was suggested by Pierre Cartier.

Some readers may have noticed that the question "what is truth in quantum mechanics?" is not mentioned in this chapter nor elsewhere in the book. The question has been however the topic of a controversy, which must be reported briefly. It originated in a paper by d'Espagnat (1989), who wondered about the relation between what was then the theory of consistent histories and the question of realism. He asked whether the word "true" can be applied to the properties occurring in a consistent history. The impossibility of using this word in most cases is due, of course, to complementarity. The present author tried to circumscribe the "true" properties, assuming that factual properties are true and proposing a criterion for extending this notion to the result of a measurement (Omnès, 1991). "True" statements were distinguished from "reliable" ones (also called "trustworthy" in a paper by d'Espagnat [1990] endorsing the notion).

More reflection or more reading should have made me aware that there can be necessary conditions for truth but no sufficient ones. This was the conclusion of a thorough analysis by Dowker and Kent (1996), who pointed out the error in my paper: the criterion I had proposed was still open to complementarity and it covered only a few cases of authenticated truth though everything about reliable properties remains correct if one wishes to use this notion.

As a matter of fact, the need for a notion of truth never occurs in interpretation when one recognizes that the purpose of a language (e.g. the one using histories) is not to decide what is real or not (i.e., making a statement about reality itself) but rather what is sensible or not (which is a statement about language). Physics needs only to make sure that it is able to account for the observed facts, and this is discussed as one of the fundamental questions in the present chapter. Looking at the use of language in the practice of physics, one can notice that nowhere does it need to make sure that the statement giving the *result* of a measurement (as opposed to a *datum* as in Chapter 21) is true in any sense of the word. The only two notions of truth that are apparently significant in physics are concerned with the facts themselves or with the validity and extent of the basic principles. For more information on this topic see d'Espagnat (1994) and Griffiths (1993).

178. The argument of this section against the soundness of the objectification problem is new. It was inspired, however, by the insights given in Griffiths (1984) and Gell-Mann and Hartle (1991), as well as in private communications. In a previous book (Omnès, 1994a), I had not yet realized the nature of the problem and it was still considered as a significant and even an essential question. This failure can certainly be attributed to an incomplete understanding of the relation between consistent histories and language, which became clearer after the criticisms by d'Espagnat (1989, 1994) and by Dowker and Kent (1996).

Chapter 21

Theorems 1 to 4 were first given in Omnès (1992), see also Omnès (1994a, Chapter 8), where proofs, more general though less simple than the ones in Chapter 19, are given.

The "Luders formula" was published by Lüders (1951).

The relation between logical and statistical probabilities is a new result. A complete proof has not been published because most of the necessary techniques and ideas can be derived immediately from a nice paper by Mittlestaedt (1997).

Chapter 22

Decoherence. The experiment we describe was performed by Brune, et al. (1996). I wish to thank Jean-Michel Raimond and Serge Haroche for useful explanations.

Entangled States

The reader might be surprised of the statement "quantum mechanics is not separable" written in Section 193, whereas an apparently contrary sentence was written in Omnès (1994a, Chapter 9). This is due to the existence of two different notions of separability: One of them (as in Section 193) refers to the conception of a physical system or its description by language (including a mathematical language). In that sense, quantum mechanics is not separable because of the existence of entangled states. Another conception of nonseparability speaks of an effect on a subsystem resulting from an action on a distant one. This is the one I rejected in a previous book.

Bell's inequalities were proposed by Bell (1964) (reproduced by Wheeler and Zurek [1983], p. 403), see also Bell (1966, p. 397), Clauser et al. (1969, p. 409). Several experiments were attempted for verifying the inequalities, and the one I report is by Aspect, Grangier, and Roger (1981); Aspect, Dalibart, and Roger (1982). Other similar tests were proposed by Horne, Shimony, and Zeilinger (1990), by Mermin (1990), and by many other people. My exposition of this topic is inspired by an article in French by A. Aspect and P. Grangier in *Le courrier du CNRS* 1984, *Images de la physique*, Supplément 55. I thank Philippe Grangier for useful comments on the experimental tests of interpretation.

Observing a Unique Atom

The experiment described here is the Dehmelt experiment: Nagourney, Dandberg, and Dehmelt (1986). See also Slauter et al. (1986); Bergquist et al. (19856). An ion trap is described by Paul (1990). About neutral atom trapping, see Cohen-Tannoudji and Dalibard (1986); King, Knight, and Wodkiewicz (1987); Zeller, Marte, and Walls (1987); Reynaud, Dalibard, and Cohen-Tannoudji (1988); Cohen-Tannoudji and Phillips (1990). The method of histories was applied to these experiments by Reynaud (1982) using the approach by Aharonov, Bergman, and Lebowitz (1964). My discussion implicitly assumes a photon localization, which is justified in Omnès (1997b).

Addendum

So many beautiful experiments have been performed in the last few years that one would like to describe most of them. The main limitation is not so much space than the author's shortcomings when seeing the subtleties and the powerful techniques they use. Putting all of them together will probably yield a good treatise on physics, and I will therefore only mention a few very incomplete references, including some remarks one can find in them.

Many older books on interpretation insist on the idea that a measurement must strongly disturb a measured system. This is not always true and quantum "nondemolition" measurements are possible. See Braginski, Vorontsov, and Khalili (1977); Braginski, Vorontsov, and Thorne (1980); Unruh (1978).

There was also a tendency in the past to identify the two notions of Macroscopic and Classical. This was questioned for the first time by Leggett (1980, 1984). The devices which had been proposed by Leggett were realized with superconducting loops containing a Josephson junction. The Hamiltonian of this system is equivalent to a particle Hamiltonian in a potential, the position of the model particle replacing the magnetic flux through the loop. This is a microstable observable as shown by the law of induction. The conjugate momentum is the voltage across the loop. One can therefore realize a macroscopic system (a SQUID) with a very narrow potential barrier allowing for macroscopically apparent tunnel effects. The effect is manifested through the sudden random production of a voltage. A lucid review is given by Clarke et al. (1988).

One should also mention:

- Interferences with two different photons: Gosh and Mandel (1987); Mandel (1983); Zou, Wang, and Mandel (1991); Wang, Zou, and Mandel (1991).
- The "quantum eraser": The data from this kind of device distinguish essentially whether a decoherence effect has occurred or not. It was proposed and discussed by Scully, Shea, and McCullen (1978); Scully, Englert, and Walther (1981); Greenberger and Sin (1989). For the experiments, see Shih and Alley (1988); Kwiat, Steinberg, and Chiao (1992); Summhammer et al. (1992).
- Delayed-choice experiments: They consist in deciding through a random choice whether a pathway detector will be active or not when the wave function of a particle is already located inside the two arms of an interferometer. If a real reduction effect had existed, the experiment would show when it occurs. The question has become mostly irrelevant but the experiments are anyway interesting as a confirmation that only objective conditions matter when a part of the device is acting. See the discussion by Wheeler in Wheeler and Zurek (1983), and in Omnès (1994a, Chapter 11). For the experiments, see Helmuth, Zajonc, and Walther (1986); Alley, Jakubowicz, and Wikes (1987).

These notes conclude a too brief review, which can only give a partial idea of a very active experimental research. I apologize to the many physicists whose work was not quoted, often because of an incomplete knowledge of a too wide domain.

References

The abbreviation WZ in some references stands for their reproduction in the source book by Wheeler and Zurek (Wheeler, 1983).

Aharonov, Y., P. G. Bergman, J. L. Lebowitz (1964) *Phys. Rev.* **134**, 1410B (WZ, p. 680).

Aharonov, Y., and D. Bohm (1961) *Phys. Rev.* **122**, 1649.

Aharonov, Y., and L. Vaidman (1991) *J. Phys.* **A24**, 2315.

Alley, C. O., O. G. Jakubowicz, W. C. Wikes (1987) In *Proc. 2nd Int. Symp. on the Foundations of Quantum Mechanics*, ed. M. Namiki. Tokyo: Physical Society of Japan, p. 36.

Anderson, P. W. (1972) *Science* **177**, 393.

Araki, H., and M. Yanabe (1960) *Phys. Rev.* **120**, 622.

Aspect, A., J. Dalibard, G. Roger (1982) *Phys. Rev. Lett.* **49**, 1804.

Aspect, A., P. Grangier, G. Roger (1981) *Phys. Rev. Lett.* **47**, 460.

Balian, R., R. Y. Alhassid, H. Reinhardt (1986) *Phys. Reports* **131**, 1.

Bargmann, V. (1954) *Ann. Math.* **59**, 1.

Bargmann, V., and E. P. Wigner (1948) *Proc. Nat. Acad. Sci.* **34**, 211.

Bell, J. S. (1964) *Physics* **1**, 195 (WZ, p. 403).

——— (1966) *Rev. Mod. Phys.* **38**, 447 (WZ, p. 397).

——— (1975) *Helv. Phys. Acta* **48**, 93.

——— (1987) *Speakable and Unspeakable in Quantum Mechanics*. Cambridge: Cambridge University Press.

Bergquist, J. C., R. B. Hulet, W. M. Itano, D. J. Wineland (1986) *Phys. Rev. Lett.* **57**, 1699.

Birkhoff, G., and J. von Neumann (1936) *Ann. Math.* **37**, 818.

Bohm, D. (1951) *Quantum Theory*. Englewood Cliffs: Prentice-Hall.

——— (1952) *Phys. Rev.* **85**, 166.

Bohm, D., and J. Bub (1966) *Rev. Mod. Phys.* **38**, 453.

Bohm, D., and B. J. Hiley (1993) *The Undivided Universe*. New York: Routledge.

Bohr, N. (1928a) In *Atti del Congresso Internationale dei Fisici, Como, 1927*. Bologne: N. Zachinelli.

——— (1928b) *Nature* **121**, 580.

——— (1934) *Atomic Physics and the Description of Nature*. Cambridge: Cambridge University Press.

——— (1935a) *Nature* **136**, 65.

——— (1935b) *Phys. Rev.* **48**, 696.

——— (1958) *Atomic Physics and Human Knowledge*. New York: Wiley.

Borel, E. (1937) *Valeur pratique et philosophie des probabilités*. Paris: Gauthier-Villars.

——— (1941) *Le jeur, la chance et les théories scientifiques modernes*. Paris: Gallimard.

Born, M. (1926) In *Quantum Theory and Measurement*, eds. J. A. Wheeler, and W. H. Zurek. Princeton: Princeton University Press.

Born, M., and E. Wolf (1959) *Principles of Optics*. London: Pergamon.

Braginski, V. B., Y. I. Vorontsov, F. I. Khalili (1977) *Zh. Eksp. Teor. Fiz.* **73**, 1340. (*Sov. Phys. JETP* **46**, 705 (1977)).

Braginski, V. B., Y. I. Vorontsov, K. P. Thorne (1980) *Science* **209**, 547.

Brune, M., E. Hagley, J. Dreyer, X. Maître, A. Maali, C. Wunderlich, J. M. Raimond, S. Haroche (1996) *Phys. Rev. Lett.* **77**, 4887.

Caldeira, A. O., and A. J. Leggett (1983a) *Physica* **A121**, 587.

——— (1983b) *Ann. Phys.* (NY) **149**, 374 and **153**, 445 (Erratum).

Caroli, C., and P. Nozières (1996) In *The Physics of Sliding Friction*, eds. B. N. J. Persson and E. J. Tosatti. Dordrecth: Klüwer.

Chevalley, C. (1997) *Ontologie et méthode dans la physique contemporaine; la physique quantique et la fin de la philosophie classique de la nature*. Thesis, University Paris X, Nanterre.

Clarke, J., A. N. Cleland, M. H. Devoret, D. Estève, J. M. Martinis (1988) *Science* **239**, 992.

Clauser, J. F., M. A. Horne, A. Shimony, R. A. Holt (1969) *Phys. Rev. Lett.* **23**, 880 (1969) (WZ, p. 409).

Cohen-Tannoudji, C., and J. Dalibard (1986) *Europhysics News* **1**, 441.

Cohen-Tannoudji, C., B. Diu, F. Laloë (1977) *Quantum Mechanics*. New York: John Wiley.

Cohen-Tannoudji, C., and W. D. Phillips (1990) *Physics Today* **43** (October).

d'Espagnat, B. (1976) *Conceptual Foundations of Quantum Mechanics*. Reading, Mass.: W. A. Benjamin.

d'Espagnat, C. (1985) *Une incertaine réalité*. Paris: Gauthier-Villars. English translation: *Reality and the Physicist*. (Cambridge: Cambridge University Press, 1989).

——— (1989) *J. Stat. Phys.* **56**, 747.

——— (1990) *Found. Phys.* **20**, 1147.

——— (1994) *Le Réel voilé, analyse des concepts quantiques*. Paris: Fayard. English translation: *Veiled Reality* Reading: Addison-Wesley, 1995.

Daneri, A., A. Loinger, G. M. Prosperi (1966) *Nuovo Cimento* **44B**, 119.

Darrigol, O. (1992) *From c-Numbers to q-Numbers: The Classical Analogy in the History of Quantum Mechanics*. Berkeley: University of California Press.

Davies, E. B. (1974) *Comm. Math. Phys.* **39**, 91.

De Witt, B. S., and N. Graham, eds. (1973) *The Many-Worlds Interpretation of Quantum Mechanics*. Princeton: Princeton University Press.

Dirac, P. A. M. (1930) *Quantum Mechanics*. Oxford: Oxford University Press.

Dowker, F., and A. Kent (1996) *J. Stat. Phys.* **82**, 1575.

Dürr, D., S. Goldstein, N. Zanghi (1992a) *Phys. Lett.* **A172**, 6.

——— (1992b) *J. Stat. Phys.* **67**, 843.

Egorov, Yu. V. (1969) *Uspehi Mat. Nauk.* **24**, 5, 235.

Ehrenfest, P., and T. Ehrenfest (1959) *The Conceptual Foundations of the Statistical Approach in Mechanics*. Ithaca, N. Y.: Cornell University Press.

Einstein, A., B. Podolsky, N. Rosen (1935) *Phys. Rev.* **47**, 777.

Emch, G. (1965) *Helv. Phys. Acta* **38**, 164.

——— (1972) *Helv. Phys. Acta* **45**, 1049.

Everett, H. (1957) *Rev. Mod. Phys.* **29**, 454.

Fefferman, C. (1983) *Bull. Amer. Math. Soc.* **9**, 129.

Feynman, R. P. (1965) *The Character of Physical Law*. London: Penguin.

Feynman, R. P., and A. R. Hibbs (1965) *Quantum Mechanics and Path Integrals*. New York: McGraw-Hill.

Feynman, R. P., R. B. Leighton, M. Sands (1965) *The Feyman Lectures on Physics*. Vol. 3. Reading, Mass. Addison-Wesley.

Feynman, R. P., and F. L. Vernon, Jr. (1963) *Ann. Phys.* (NY) **24**, 118.

Foldy, L. L., and S. A. Wouthuysen (1950) *Phys. Rev.* **78**, 29.

Gell-Mann, M. (1994) In *Proceedings of the 4th Drexel Symposium on Quantum Non-Intehrability: The Quantum-Classical Correspondence*, Drexel University, Philadelphia, Pa. September 8-11, 1994, D. H. Feng, ed., International Press, 1996.

Gell-Mann, M., and J. B. Hartle (1991) In *Complexity, Entropy, and the Physics of Information*, ed. W. H. Zurek. Redwood City, Calif.: Addison-Wesley.

——— (1993) *Phys. Rev.* **D47**, 3345.

Ghirardi, G. C., A. Rimini, T. Weber (1986) *Phys. Rev.* **D34**, 470.

Ginibre, J., and G. Velo (1979) *Comm. Math. Phys.* **66**, 37.

Gisin, N., and I. C. Percival (1992) *J. Phys.* **A25**, 5677.

——— (1993) *J. Phys.* **A26**, 2245.

Giulini, D., E. Joos, C. Kiefer, J. Kupsch, O. Stamatescu, H. D. Zeh (1996) *Decoherence and the Appearance of a Classical World in Quantum Theory*. Berlin: Springer.

Gleason, A. M. (1953) *J. Math. Mechanics* **6**, 895.

Gosh, R., and L. Mandel (1987) *Phys. Rev. Lett.* **59**, 1903.

Greenberger, D. M., and A. Ya. Sin (1989) *Found. Phys.* **19**, 679.

Griffiths, R. G. (1984) *J. Stat. Phys.* **36**, 219.

——— (1987) *Am. J. Phys.* **55**, 11.

——— (1993) *Found. Phys.* **23**, 1601.

——— (1986) *Phys. Rev.* **A54**, 2759.

——— (1998) *Phys. Rev.* **A57**, 1604.

Grigis, A., and J. Sjöstrand (1994) *Microlocal Analysis for Differential Operators*. London Mathematical Society Lecture Note Series 196. Cambridge: Cambridge University Press.

Haag, R., and D. Kastler (1964) *J. Math. Phys.* **77**, 1.

Haake, F. (1973) *Springer Tracts in Modern Physics* **66**, 98.

Hagedorn, G. (1980a) *Comm. Math. Phys.* **71**, 77.

——— (1980b) *Comm. Math. Phys.* **77**, 1.

——— (1981) *Ann. Phys.* (N.Y.) **135**, 58.

Halliwell, J. J., and A. Zoupas (1995) *Phys. Rev.* **D52**, 7294.

Haroche, S., and J. M. Raimond (1994) In *Advances in Atomic, Molecular, and Optical Physics. Supplement I*, ed. P. Berman. New York: Academic Press, p. 123.

Heisenberg, W. (1930) *Die physikalischen Prinzipien der Quantentheorie*, Springer. English translation: *The Physical Principles of the Quantum Theory*. Chicago: University of Chicago Press, Trans. Carl Eckart and Frank C. Hoyt.

——— (1958) *Physics and Philosophy: The Revolution in Modern Science*. New York: Harper & Row.

Helmuth, T., A. G. Zajonc, H. Walther (1986) In *New Techniques and Ideas in Quantum Measurement Theory*, ed. D. M. Greenberger, Proc. New York Acad. Sci., **480**, 108.

Hepp, K. (1972) *Helv. Phys. Acta* **45**, 237.

——— (1974) *Comm. Math. Phys.* **35**, 265.

Hepp, K., and E. H. Lieb (1974) *Helv. Phys. Acta* **46**, 573.

Hörmander, L. (1979a) *Comm. Pure Appl. Math.* **32**, 359.

——— (1979b) *Ark. Mat.* (Sweden) **17**, 297.

——— (1985) *The Analysis of Partial Differential Operators*. 3 vols. Berlin: Springer.

Horne, M. A., A. Shimony, A. Zeilinger (1990a) *Phys. Rev. Letters* **62**, 2209.

——— (1990b) *Nature* **347**, 429.

Hu, B. L., J. P. Paz, Y. Zhang (1992) *Phys. Rev.* **D45**, 2843.

Isham, C. J., and N. Linden (1994) *J. Math. Phys.* **35**, 5452.

——— (1995) *J. Math. Phys.* **36**, 5392.

Isham, C. J., N. Linden, S. Schrekenberg (1994) *J. Math. Phys.* **35**, 6360.

Jaklevic, R. C., J. Lambe, A. H. Silver, J. E. Mercereau (1964) *Phys. Rev. Lett.* **12**, 159.

Jammer, M. (1966) *The Conceptual Development of Quantum Mechanics*. New York: McGraw-Hill.

——— (1974) *The Philosophy of Quantum Mechanics*. New York: Wiley.

Jauch, J. M. (1964) *Helv. Phys. Acta* **37**, 293.

Joos, E., and H. D. Zeh (1985) *Z. Phys.* **B59**, 229.

Jost, R. (1976) *Studies in Mathematical Physics: Essays in Honor of Valentine Bargmann*, eds. E. H. Lieb, B. Simon, and A. S. Wightman. Princeton Series in Physics, Princeton: Princeton University Press.

Karolyhazy, F. (1966) *Nuovo. Cim.* **A42**, 390.

——— (1974) *Magyar Fizikai Polyoir Mat.* **12**, 24.

Karolyhazy, F., A. Frenkel, B. Lukacs (1986) In *Quantum Concepts in Space and Time*, eds. R. Penrose and C. J. Isham. Oxford: Oxford University Press, pp. 109–28.

Kent, A. (1997) *Phys. Rev. Lett.* **78**, 2874.

King, M. S., P. L. Knight, K. Wodkiewicz (1987) *Opt. Comm.* **62**, 385.

Kwiat, P. G., A. M. Steinberg, R. Y. Chiao (1992) *Phys. Rev.* **A45**, 7729.

Landau, L. D., and E. M. Lifschitz (1967) *Statistical Mechanics*. Oxford: Pergamon.

——— (1968) *Quantum Mechanics*. Oxford: Pergamon.

Leggett, A. J. (1980) *Progr. Theor. Phys., Supplement* **69**, 1.

——— (1984) *Phys. Rev.* **B30**, 1208.

——— (1987) In *Chance and Matter*, Les Houches Session XLVI, 1986, eds. J. Souletie, J. Vannimenus, R. Stora, Amsterdam: North Holland.

Lévy-Leblond, J. M. (1963) *J. Math. Phys.* **4**, 726.

Lindblad, G. (1979) *Comm. Math. Phys.* **65**, 281.

London, F., and E. Bauer (1939) *La théorie de l'observation en mécanique quantique*. Paris: Hermann.

Lüders, G. (1951) *Ann. Phys.* **8**, 322.

Maeda, S. (1990) *Reviews in Mathematical Physics* **1**, 235.

Mandel, L. (1983) *Phys. Rev.* **A28**, 929.

Mehra, J. (1973) *The Physicist's Conception of Nature*. Dordrecht: Reidel.

Mermin, N. D. (1990) *Am. J. Phys.* **58**, 731.

Merzbacher, E. (1970) *Quantum Mechanics*. New York: John Wiley.

Messiah, A. (1995) *Quantum Mechanics*. Amsterdam: North-Holland.

Mittlestaedt, P. (1978) *Quantum Logic*. Dordrecht: Reidel.

——— (1997) In *New Developments on Fundamental Problems in Quantum Physics*, eds. M. Ferrero and van der Merwe. Dordrecht: Kluwer, p. 265.

Mott, N. (1929a) *Proc. Roy. Soc.* **A126**, 79.

——— (1929b) *Proc. Roy. Soc.* **A124**, 440.

Murdoch, D. (1987) *Niels Bohr's Philosophy of Physics*. Cambridge: Cambridge University Press.

Nagourney, W., J. Dandberg, A. Dehmelt (1986) *Phys. Rev. Lett.* **57**, 384.

Naimark, M. (1959) *Normed Rings*. Gröningen: P. Nordhoff.

Nakajima, S. (1958) *Prog. Theor. Phys.* **20**, 948.

Newton, T. D., and E. P. Wigner (1949) *Rev. Mod. Phys.* **21**, 400.

Omnès, R. (1988) *J. Stat. Phys.* **53**, 893.

——— (1989) *J. Stat. Phys.* **57**, 357.

——— (1990) *Ann. Phys.* (N.Y.) **201**, 354.

——— (1991) *J. Stat. Phys.* **62**, 841.

——— (1992) *Rev. Mod. Phys.* **64**, 339.

——— (1994a) *The Interpretation of Quantum Mechanics*. Princeton: Princeton University Press.

——— (1994b) *Philosophie de la science contemporaine*. Paris: Gallimard. Trans. by A. Sangalli (with additional material). *Quantum Philosophy*. Princeton: Princeton University Press, 1999.

——— (1997a) *J. Math. Phys.* **38**, 697.

——— (1997b) *J. Math. Phys.* **38**, 708.

——— (1997c) *Phys. Rev.* **A56**, 3383.

Paul, W. (1990) *Rev. Mod. Phys.* **62**, 531.

Pauli, W. (1933) Die allgemeinen Prinzipien der Wellenmechanik. In *Handbuch der Physik*, eds. H. Geiger and K. Scheel, Springer, Berlin, vol. 24. Reedited in *Handbuch der Physik* (*Encyclopaedia of Physics*), Springer (1958), vol. 5.

——— (1936) *Scientia* **59**, 65.

——— (1950) *Experientia* **6**, 72.

——— (1979) In *Wissenschaftlicher Briefwechsel mit Bohr, Einstein, Heisenberg, u.a.*, eds. K. von Megenn, A. Hermann, V. Weisskopf. 3 vols. Berlin: Springer.

Paz, J. P., S. Habib, W. H. Zurek, *Phys. Rev.* **D47**, 488.

Paz, J. P. (1994) In *Physical Origins of Time Asymmetry*, eds. J. J. Halliwell, J. Pérez-Mercader, W. H. Zurek. Cambridge: Cambridge University Press.

Pearle, P. (1976) *Phys. Rev.* **D13**, 857.

——— (1984) *Phys. Rev. Lett.* **53**, 1775.

——— (1989) *Phys. Rev.* **A39**, 2277.

Penrose, R. (1996) *Gen. Rel. Grav.* **28**, 581.

——— (1997) *The Large, the Small, and the Human Mind.* Cambridge: Cambridge university Press.

Peres, A. (1980) *Phys. Rev.* **D22**, 879.

——— (1995) *Quantum Theory: Concepts and Methods.* Dordrecht: Kluwer.

Pfeifer, P., and J. Fröhlich (1996) *Rev. Mod. Phys.* **67**, 759.

Primas, H. (1981) *Chemistry, Quantum Mechanics, and Reductionism.* Springer, Berlin.

Pryce, M. H. L. (1948) *Proc. Roy. Soc.* (London) **A195**, 62.

Reed, M., and B. Simon (1972) *Methods of Modern Mathematical Physics.* New York: Academic Press.

Reynaud, S. (1982) *Ann. Phys.* (Paris) **8**, 315.

Reynaud, S., J. Dalibard, C. Cohen-Tannoudji (1988) *IEEE J. Quant. Electr.* **24**, 1395.

Rosenthal, S., ed. (1967) *Niels Bohr: His Life and Work as Seen by His Friends and Colleagues.* Amsterdam: North-Holland.

Scheck, R., T. Brun, I. C. Percival (1995) *J. Phys.* **A28**, 5041.

Schilpp, P. A., ed. (1949) *Albert Einstein, Philospher-Scientist.* Evanston, Ill.: Library of Living Philosophers.

Schrödinger, E. (1935) *Naturwissenschaften* **23**, 807, 823, 844.

Schwartz, L. (1950) *Théorie des distributions.* Paris: Hermann.

Scully, R. O., B-G. Englert, H. Walther (1981) *Nature* **351**, 111.

Scully, R. O., R. Shea, J. D. McCullen, (1978) *Phys. Rev.* **43**, 485.

Selleri, F. (1990) *Quantum Paradoxes and Physical Reality.* Dordrecht: Klüwer.

Sewell, G. L. (1967) *Physica* **34**, 493.

Shih, Y. H., and C. O. Alley (1988) *Phys. Rev. Lett.* **61**, 2921.

Slauter, T., W. Neuhauser, R. Blatt, P. E. Toschek (1986) *Phys. Rev. Lett.* **57**, 1696.

Stein, E. M. (1993) *Harmonic Analysis: Real-Variable Methods, Orthogonality, and Oscillatory Integrals.* Princeton: Princeton University Press.

Summhammer, J., G. Badurek, H. Rauch, U. Kischic (1992) *Phys. Lett.* **A90**, 110.

Stern, O. and W. Gerlach (1924) *Zeitschrift für Physik* **74**, 673.

Tanguy, A., and P. Nozières (1996) *J. Physique I* **6**, 1251.

Thirring, W. (1979) *Quantum Mechanics of Atoms and Molecules*, p. 209 (Volume 3 de, *A Course in Theoretical physics*). Berlin: Springer.

Unruh, W. G. (1978) *Phys. Rev.* **D18**, 1764.
Unruh, W. G., and W. H. Zurek (1989) *Phys. Rev.* **D40**, 1071.
Vigier, J. P. (1956) *Structure des microobjets dans l'interprétation causale de la théorie des quanta*. Paris: Gauthier-Villars.
van Kampen, N. G. (1954) *Physica* **20**, 603.
van Heijenoort, J. (1967) *From Frege to Gödel: A Source Book in Mathematical Logic*. Cambridge: Harvard Press.
von Neumann, J. (1927) *Math. Zeitschr.* **26**, 1.
——— (1932) *Mathematische Grundlagen der Quantenmechanik.* Berlin: Springer. Trans. R. T. Beyer, *Mathematiical Foundations of Quantum Mechanics*. Princeton University Press (1955).
——— (1940) *Ann. of Math.* **41**(2), 94.
Voros, A. (1978) *J. Func. Analysis* **29**, 104.
Wang, L. J., X. Y. Zou, L. Mandel (1991) *Phys. Rev.* **A44**, 4614.
Weinberg, S. (1995) *The Quantum Theory of Fields*. Cambridge: Cambridge University Press.
Weyl, H. (1950) *Bull. Amer. Math. Soc.* **56**, 115.
Wheeler, J. A., and W. H. Zurek (1983) *Quantum Theory and Measurement.* Princeton: Princeton University Press.
Whitten-Wolfe, B., and G. Emch (1976) *Helv. Phys. Acta* **49**, 45.
Wick, G. C., A. S. Wightman, E. P. Wigner (1952) *Phys. Rev.* **88**, 101.
Wightman, A. S. (1962) *Rev. Mod. Phys.* **34**, 845.
Wigner, E. P. (1932) *Phys. Rev.* **40**, 749.
——— (1939) *Ann. Math.* **40**, 149.
——— (1952) *Zeitschr. Phys.* **131**, 101.
——— (1959) *Group Theory and Its Application to Atomic Spectra*. New York: Academic Press.
——— (1967) *Symmetries and Reflections*. Bloomington: Indiana University Press.
——— (1976) Interpretation of quantum mechanics. In WZ, p. 260.
Zeh, H. D. (1970) *Found. Phys.* **1**, 69 (WZ, p. 342).
Zeller, P., M. Marte, D. F. Walls (1987) *Phys. Rev.* **A35**, 198.
Zou, X. Y., L. J. Wang, L. Mandel (1991) *Phys. Rev. Lett.* **67**, 318.
Zurek, W. H. (1981) *Phys. Rev.* **D24**, 1516.
——— (1982) *Phys. Rev.* **D26**, 1862.
——— (1991) *Physics Today* **44**(10), 36.
——— (1994) In *Physical Origins of Time Asymmetry*, eds. J. J. Halliwell, J. Pérez-Mercader, W. H. Zurek. Cambridge: Cambridge University Press.
Zwanzig, R. (1960) *Lect. Theor. Phys.* (Boulder) **3**, 106.
——— (1964) *Physica* **30**, 1109.

Index